Boss Snakes

Burmese python,
an introduced species in the Everglades.

Boss Snakes

Stories and Sightings of Giant Snakes in North America

Coachwhip Publications
Landisville, Pennsylvania

Contents

Acknowledgments

Many thanks to Gary Mangiacopra, Craig Heinselman, Loren Coleman, and Rod Dyke for passing along snake stories. Thanks, also, to the Postmaster of Luray, Kansas, and the private collector she located, for a sample of the special cancellation celebrating Luray's giant snake.

Snake Stories.

There has been a dearth of snake stories since the war. Hence, it was exciting to read the other day about the monster serpent of Salem, Mich. A posse of well-armed citizens had been hunting the creature. It was said to measure 17 feet, to outrun a man. Farmers have reported it off and on for the past year.

Snakes were not always the sluggish creatures they have become of late. A big snake was reported recurrently for 30 years in the neighborhood of Easton, Md. It was alleged to be as big around as a saucer, and 15 to 20 feet long. When last hunted by an organized "snake posse" it was seen tearing through the woods, its head raised four feet from the ground and making as much noise as a horse.

Hoop snakes formerly were wont to roll over Tennessee hills, burying the spikes of their venomous tails in tall oaks, withering the trees like strokes of lightning. From Atlanta came details of a terrible struggle between an aged farmer and a giant rattlesnake, in which the vengeful reptile followed the farmer for miles.

From the bad lands of North Dakota issued the story of a hunter who was amusing himself as he sat on a rock playing a harmonica when he suddenly realized that his tunes had attracted hundreds of rattlesnakes. As long as he played they swayed dreamily to the rhythm, but as soon as he stopped they became angry and coiled to strike. He was saved by a companion who set fire to the prairie grass.

Postwar snake news has been rather routine. Snake prices were quoted from South Africa— puff adders at \$1.50 a foot and cobras at \$1.70, and it was revealed that a price war between retailers in Chicago has reduced the rate on live snakes to 18 cents a pound.

There have been a number of two-headed snakes, one a rattler in West Virginia, one a garter snake in Oregon. In Texas a snake crawled into a transformer and caused a short circuit that kept the city of Cuero in darkness for half an hour, and in Topeka a bull snake had a like exploit.

The British call summer the silly season. It is traditionally the time for sea serpents to raise their grisly heads above the waves and for land serpents to show their stuff. May the little incident in Salem presage a fruitful summer.

—Editorial, July 1, 1949, Saint Joseph, Michigan, *Herald Press*.

This same editorial was reprinted or excerpted in various newspapers over the next several decades, at least through 1965.

Introduction

In 1995, the *INFO Journal* (International Fortean Organization) published a short historical piece I wrote on giant snake stories from Harford and Baltimore Counties, Maryland. Since that time, I have continued to collect reports of oversized ophidians throughout North America. My previous book, *Cryptozoology: Science & Speculation*, included stories of the Broad Top serpent here in Pennsylvania, and giant snake tales from the Everglades, but I have always wanted to put together a specific volume collecting our indigenous big snake folklore.

My continuing interest is both cryptozoological and herpetological. I started investigating mystery animals in the late 1980s, and not long after that, my wildlife interests focused on reptiles and amphibians, particularly on colubrid snakes. I've been called a critic of the giant snake stories (Coleman 2001), but more truthfully, I'm critical of the too-often sloppy evaluation of such accounts.

I anticipate some confusion by reptile enthusiasts who encounter this book, and wonder what's going on. I can't possibly be proposing that there are unknown species of giant snakes roaming the country-side, right? And what's so interesting about one old snake story after another? We all know that big snakes getting loose (or being released by irresponsible owners) are common news-talk today, and it isn't difficult to wave our hands and dismiss old sightings of big serpents as tall tales, pioneer jokes, and the occasional escape of a boa or python from a traveling circus. Often, there's good reason to do so.

So, let me quickly explain the purpose of this book. This is an attempt to bring a "broad view" picture to cryptozoological research into big snake stories of North America. Cryptozoology is the investigation of mystery animals—animals which have attracted enough attention to be reported or ethnozoologically described, but which haven't been scientifically confirmed due to lack of evidence. Obviously, not every case of a mystery animal, or *cryptid*, has an unknown species behind it. Cultural influences, exaggeration, misidentification, poor eyewitness conditions, and other factors contribute to unreliable personas of strange creatures. The purpose of cryptozoology, though, is to investigate, no matter what results are finally determined.

Giant snake stories in North America shared an early folkloric landscape with many beasts. Predators and "varmints" were deemed dangerous to health and livelihood. Worse yet were creatures more often rumored than encountered, that seemed removed from their native element, or that had strange physical or behavioral characteristics. Snakes, of course, have been treated as pests for generations, and even harmless species were (and are) killed with little thought and no regrets. It should be no surprise, then, that when stories of giant snakes were reported, they often created local panics and calls to arms. It would be poor investigative practice, however, to dismiss all accounts out-of-hand as just bogies and scare tactics, without bothering to take a closer look.

When I published *The Historical Bigfoot*, I included a classificatory scheme of possible explanations for Wild Man stories. Here, too, there are multiple potential explanations for consideration.

1. False Stories
    a. Newspaper Hoaxes: to fill up space, entertain readers, or spark sales
    b. Pranks
    c. Misidentifications
    d. Campfire Tales: stories by woodsmen, may or may not be true, but often exaggerated
    e. Rumors: spread through towns, sometimes picked up by newspapers
    f. Showman's Hype: stories to attract crowds to an exhibit
    g. Social Control: stories spread to scare and manipulate people (including property deterrents)
2. Recognized Snakes
    a. Native Species: exaggerated size
    b. Native Species: exceeding recognized size
    c. Exotic Species: out-of-place escapees, etc.
3. Unknown Snakes

The following brief chapters will introduce these topics further. But, first, I suppose, a question is begged. What exactly is a *boss* snake? There are kingsnakes and ratsnakes, milk snakes and gopher snakes... among many others. You won't find a boss snake in your herpetological field guide, though. That is a name, about a century in disuse, for any large snake that dwarfed its scaly brethren. The possibility that such snakes might exist today should be examined critically and carefully, without dubious argumentation, but examined nonetheless.

Hoaxes, Tall Tales, and Paranormalism

It should come as no surprise that fictitious tales and humbugs are quite common in the history of snake stories. Deliberate hoaxes are plentiful, and as will be seen in the state survey section, are not always easy to identify. Hoaxing must always be considered as a possible explanation in an investigation. Still, there are other ways for snake fiction to circulate. Mistaken identities are not unusual, and it is very easy to overestimate the length of a snake (particularly if you are afraid of snakes). The following stories are just a few examples of the different categories of mistaken or fictional accounts of giant snakes.

Deliberate Hoaxes and Pranks

Nancy McDaniel Webb (2002) noted that her grandfather grew up near Mt. Pleasant, Alabama. She stated, "He told me about dragging a water-melon held by the still attached vine, down the dirt road, weaving back & forth, to make a trail looking like a snake track. Of course, the 'giant snake' got the community in an uproar. He apparently got the greatest kick out of it at the time, then again every time he re-told the tale to the grandkids thereafter."

Editor's Notebook: Speaking of Monsters,
Did You Ever Hear of The Oneida Python?
by John Torinus

Last week's stories about the hunt for a monster near the Appleton sewage disposal plant and the amusing end of the

The *Horn Snake*—this and the hoop snake were the
mythological serpents most commonly popularized in
colonial America. (Illustration from Skitt (1859).) There is
a tendency to refer all horn snake stories to the mud snake
(*Farancia abacura*), because of the mud snake's sharp (but
harmless) tail. While there is folkloric integration with
Farancia, the horn snake stories are more likely an amalgamation of diverse folkloric sources.

episode recalled a hilarious story I worked on as a cub reporter out Oneida way some 25 years ago.

For a week or so there were rumors that various Oneida residents had seen a huge snake in a marsh area east of that community. It had tentatively been identified as a python. Now a python is a rather rare creature to be roaming loose in Wisconsin so over a period of several afternoons, I hied myself out there to interview people who claimed to have seen the snake. Some of them were quite lucid in their descriptions, and their comments made interesting news for about a week. There were even several small-scale posses organized to hunt down the reptile.

The No. 1 fire station in Green Bay for some reason became headquarters for the hunt and the "news" came out of that headquarters. One morning when I had just sleepily reported for work I got a call from the station to come right over, the python had been captured.

There was a rather dark stair well going down into the basement of the station and sure enough in the half light at the bottom of the well there appeared to be a giant snake. The firemen were having a great time with their exhibit until someone turned a spotlight on the reptile and found it was a fire hose which the boys had elaborately painted to resemble a python. But it made a good picture for the paper, and for a time the firemen's hoax put an end to the story.

A week or so later, however, I got a call from a tavern-keeper west of the city towards Oneida who said that for sure this time the reptile had been captured and they had it in a cage in a garage behind the tavern. I hotfooted it out with a photographer and sure enough there in the garage was a caged python about eight feet long. The tavern-keeper and a buddy told how they had captured it in the swamp east of Oneida.

I returned to the office with pictures and story and with my chest out several feet because I had been taking quite a ribbing about my snake stories. Now I had the proof and I wrote a very colorful and detailed account of the capture.

But when I turned it in to the city editor, a veteran newspaperman, he looked it over carefully while I grew more and more uneasy standing in front of his desk.

"Call the Railroad Express," he ordered, "and see if a snake has been shipped in to Green Bay recently."

I objected. I didn't see what good that would do. The sources of my story I considered infallible. But he carefully explained that he just wanted to make sure; that the only way a snake could be shipped in was by Railway Express.

Imagine my surprise when I called the local agent.

"Sure," he said, "we received a large python yesterday."

He gave me the name of the person it had been shipped to. It was my tavern-keeper friend.

When the story came out in that day's paper it was quite a different one from what the tavern-keeper had expected.

His hoax was completely exposed.

All of which is just an illustration of what a city editor on a newspaper is there for, and how a cub reporter learns his trade.

Appleton, Wisconsin, *Post-Crescent*
September 8, 1963

Now, that last story is a good reminder that even if a sighting is legitimate, publicity creates incentive for pranksters to join in, and it isn't uncommon for exposure of later hoaxes to overshadow earlier reports. This creates a situation where the entire subject becomes a joke in the public's eye. This is one reason hoaxes and mistakes need to be rooted out as soon as possible in an investigation.

The Great Russellville Snake.

A friend has communicated to us a snake story, which we think worth recording. In July last, there was a "great excitement" in and about Russellville, in this State, on account of the reported appearance in that neighborhood of an immense snake, which had been seen at different times by various credible witnesses.— The favorite haunts of his snakeship were a pond and brier patch some two miles from Russellville, on a farm of an old gentleman, who was much annoyed by visitants in quest of fish and blackberries.—Various and contradictory reports were related as to the description, size, &c., of the monster; some giving it as their opinion that it was the veritable sea serpent, which was making a tour of the continent, while others were sure it was a dry land reptile of the rattle breed, as they had heard the rattle of its mighty tail hundreds of yards off. Some contended that it was as long as a clothesline, and as big as a yearling calf, while others of a less vivid imagination represented it as not more than sixteen feet long and three feet in circumference.

A hunter had gone in pursuit of him and returned without his dog, whom it was supposed snakey had encountered and swallowed raw. A young negro had been missed from the neighborhood, and it was firmly believed he had "followed in the footsteps" of pointer, and had been gulped down, wool, boots, and all.

As might be supposed, these rumors spread with rapidity, and created terror in the minds of Russellvillians; they declared a general crusade against the serpentine intruder, and resolved to "bead him or die." Traps were set, and snares laid to catch him, and armed hunters sought him, but in vain. At length one day two negroes, who had been in search of this monster, discovered him lying across a fence near his favorite blackberry patch, which he seemed to watch with as much interest as if he was the owner thereof. One of the darkies, feeling probably that "familiarity breeds contempt," kept a long way off, and raising his gun, "fired and fell back." The other having more nerve, approached within some twenty yards and let off his double-barrel, which took effect in the reptile's head.—They immediately put off to town to relate their achievement, and procure a horse and cart to carry the carcass away. Sam proceeded at once to the most popular hotel, and to a gaping auditory proclaimed his victory over the terror of the neighborhood; and reception, as the novelist says, "may be better imagined than described." Wellington, fresh from Waterloo, of Hyer, after his fight with Sullivan, was no more lionized than Sam was at the Russellville hotel. He invigorated the internal man with frequent libations, and then, with a horse and cart, followed by numerous inhabitants, on foot and on horseback, proceeded in quest of the spoils of victory. Sam reached the place first, and uncoiled the snake from the fence— citizens then gathered around, and began a minute examination of the monster. About this time a horrid oath burst from Sam—a general roar of laughter arose form the crowd—and the "tempest in a teapot" had exploded. The wonderful and terrifying monster was made of striped muslin, stuffed with bran and shavings, and painted to the life.

The counterfeit snake was an ingenious piece of workmanship, and reflected credit upon the mechanical ingenuity of the owner of the land, whose receipts from fish and blackberries last summer, exceeded any previous year. Sam still declares that the sarpint is not the one he killed, but that a swap was made during his absence.

Madison, Wisconsin, *Express*
May 1, 1849

The Great Snake Story

The Wichita Falls *Record-News* reports that the gigantic reptile which kept the people of that community on tenterhooks for weeks has finally been exposed as the hoax of a group of teenage youths in search of a little good clean fun.

It was only a dummy, made of tow-sacks sewed together and filled with cottonseed, the whole painted black to look businesslike. It made numerous appearances along the highways and byways of the vicinity, and scared a good many people out of their wits.

Posses were organized to hunt down the monster. One man is reported to have offered $1,000 for the snake alive. Residents were made jittery by the wild rumors floating about, and the whole countryside was disturbed.

The Wichita paper says the boys had no intention of creating such excitement. They merely thought it would be a good joke. Their identity, the paper says, may never be known. The dummy has been destroyed. The show is over.

Hoaxes of this nature, it seems to us, are rarer nowadays than in the good old days. This particular hoax was different from most, in that there actually was a dummy snake, whereas most hoaxes are born and thrive on nothing more substantial than idle rumor.

The snake hoax was funny to its perpetrators, but hardly a laughing matter to those who were victimized by it. Fortunately none of the victims died of heart failure, as they might well have.

Those who were actually exposed to a sight of the snake probably are no more resentful than those who were duped into offering rewards for its capture, said to have reached a total of $5,000, or to the owners of "snake dogs" which were offered as trackers-down of the putative python.

At any rate, the big snake scare around Wichita Falls is over—unless there are some who persist in believing that the story of the hoax is itself a hoax. Error sometimes persists in the face of facts.

Abilene, Texas, *Reporter-News*
July 17, 1946

Misidentifications

Water Serpent Only Fire Hose

Montreal, Aug. 26 (UP)—The mystery surrounding the strange reptile which was reported to have been seen in a pond in a local stone quarry has been cleared up, and as a result Alderman Dubreuil is $10 richer.

The "monster," Alderman Dubreuil said, consisted of a length of fire hose controlled by strings in the hands of two lads who were trying to frighten the watchman at the quarry.

Sunk just below the surface of the pond, the hose-length assumed the wriggles of a huge reptile when the controlling cords were jerked.

Dubreuil's $10 was the result of a bet with another member of the city council, who said Dubreuil could not explain the mystery.

Port Arthur, Texas, *News*
August 26, 1938

The Bedford *News* tells the following snake story: "The biggest snake story of the season is told by Mr. Nat Williams, who says he was driving along the road some distance from town a few days ago, when he discovered a black snake some ten or twelve feet long, without any head. His curiosity was aroused, and he got out to examine it, when he found that a large snake was trying to swallow a smaller one, and had succeeded in getting two feet of it out of sight."

Fort Wayne, Indiana, *Daily Gazette*
August 10, 1870

Town Turns Out to Fight Reptile
Ohioans Show Bravery in Defending Homes,
Although Battle Becomes Fiasco

Logan, O.— "Help, help! A snake 40 feet long is chasing me!"
Shouting between gasps for breath, Gerald Grimmer dashed
from the woods on Snow Fork.

"It's as big around as my waist and its right down by the edge
of the pond, curled around a tree," cried the boy.

The village was aroused. A monster reptile imperiled the live
stock, not to mention small boys. Shotguns, pitchforks, scythes,
any weapons that came to hand were grasped as the fire bells
called the village to fight the slimy invader.

Two hundred men advanced upon the woods, weapons held
ready. Scouts were thrown out, while the pseudo army deployed
as skirmishers.

"Hiss-s-s-s-s!"

The attackers retreated. A council of war was held.

The honor and safety of Snow Fork were at stake. Some brave
spirit gave the word to advance and the battle was on.

Step by step the champions of local pride closed in on the
foe. A shot echoed through the woods. Another. Yet another.

"I got him!" shouted a villager, as a load of buckshot tore into
the dimly outlined form on the tree.

An angry hiss, a thrashing in the branches, and the writhing
form dropped to the ground, quivered spasmodically, and was still.

The men of Snow Fork had saved the day.

Five years ago, Jesse Thompson, a sawyer, had a mill on Snow
Fork. To siphon water from the pond he used a 40-foot length of
hose. Later he left the woods, also the hose.

Now the hose is unfit for use, being perforated by buckshot.

The hisses? A swarm of bees which occupied the hose objected
to acting as targets.

> Waynesboro, Pennsylvania, *Press*
> January 24, 1921

We understand that a party of adventurous young men of
Cambridge recently went out to the snake hole on the Frazier farm

and after digging about thirty feet, unearthed a groundhog, the elongation of which in his precipitate flight from an intruder appeared like a monster serpent. The snake still lives and will appear again next dog days.

Cambridge, Ohio, *Jeffersonian*
August 17, 1893

It Is Said—

That the mystery of the mythical monster snake which was reported making gigantic tracks in New London gardens last week and stirring up the imaginations of townsfolk as to what it might be, was cleared up this week in a letter written by a 10-year-old boy, Dick Schwan, son of Mr. and Mrs. Herbert Schwan, 1314 S. Pearl street, to the local newspaper. The letter went as follows:

"Dear People:
"Don't be afraid of monsterous snakes any more. I guess it was only my dog that broke loose and dragged his chain. I was away and he got lonesome and looked for someone to play with. He gets so lonesome when I am away. I had his ten foot chain on a big spike but he got loose anyway. The chain was only one-half inch wide. I am very sorry if he went through gardens. I looked and looked for him but I couldn't find him. This is the first time he has been loose since April 1st.
"My dog climbs the ladder and goes down the chutes at school. He wouldn't hurt anyone. I won't let him get away again.
"Dick Schwan."

Appleton, Wisconsin, *Post-Crescent*
July 21, 1939

Emmons County's Snake 'Captured'

Linton, N, D., June 21.—The mystery of Emmons county's giant snake that left a three-inch deep trail as it wriggled over the prairie has been solved. Harry Nagel of North Winona reports the monster was a bull, not a bullsnake. Attached to the bull's collar was a pole 16-feet long. The contrivance kept the bull from exercising his habit of jumping fences.

Bismarck, North Dakota, *Tribune*
June 21, 1935

Exaggeration

Snake Story.

Evan Williams is authority for the statement that on the Vol. Mozier farm, 7 miles south of Delphos, the other day, a snake was seen that measured fully fifteen feet in length. The monster took refuge in a tile ditch and escaped its would be slayers.

Delphos, Ohio, *Daily Herald*
June 18, 1897

Evan's Snake Story.

The snake story told by Evan Williams stirred up a number of Needmore people. Henry Mosier, Albert Mosier, Mr. Poling and the pioneer hunter Henry Gallaspie excavated several rods of tile ditch in the search for the monster before they began to wonder how a snake 10 inches in diameter could crawl through an 8-inch tile ditch. It is needless to say that the snake is still at large.

Delphos, Ohio, *Daily Herald*
June 23, 1897

Snake Hunt

Alexandria, La. (UPI)—Deputies and firemen hurried to a field in five police cars and a fire truck Friday to search for a monster snake reported seen in the area.

Service station operator Jim A. Vallet and a high school girl told authorities they saw a 15-foot snake slither out of a hay wagon.

Deputies set fire to the field. Then the firemen took over and trained high-pressure hoses into culverts.

A snake that wanted only to be left alone floated out of a culvert and was beaten to death by deputies and firemen.

But the battered body measured only 59 inches.

"I guess he must have shrunk when he saw all the commotion." Vallet said.

Lowell, Massachusetts, *Sun*
October 9, 1966

Tall Tales

The black snake, while not commonly feared for its bite, is nevertheless a dangerous serpent, belonging, as it does, to the constrictor family, which chokes or crushes his victims to death. Instances are not lacking hereabout of their having exercised their peculiar powers very often seriously, and, it is said, sometimes fatally. There are numerous cases of black snakes attacking people and twining about them. In one case a girl was said to have been choked to death, and in another case a big snake nearly killed a man, and probably would have done so had he not succeeded in getting his knife out of his pocket and cutting the reptile to pieces.

A gentleman who was talking on this subject a few days ago told of the fate of a hunter along the Wallkill. George Wisner had gone hunting with a friend in the vicinity of Merritt's island, a few miles from Unionville. The men were in canoes. The friend saw Wisner shoot some ducks, and following up his shot, disappear behind a clump of willows, around a bend. When it came time to return the friend tried to find Wisner. He searched the neighborhood until nearly dark, when he discovered the canoe drawn up along shore. He shouted and searched without avail, and finally gave up and sent an alarm. A party was organized and the search kept up all night with lanterns. Nest morning Wisner's lifeless body was found. A great black circle extended around the body and another around the neck. The soft ground was greatly trampled, as though a struggle had taken place. Leading from the body was the sinuous marks where a gigantic snake had crawled away. It was believed that Wisner had been killed by a monster black snake which, in those days, was said to inhabit that part of the drowned lands, but whether the story of this great snake was an outgrowth of the fearful death of the hunter, or the death of the hunter developed from the big snake story, is a matter of conjecture.

Galveston, Texas, *Daily News*
July 4, 1885

The Snake Liked Persimmons

Recently Madison Jolly, a well known negro of Green county, was attacked by a large rattle snake about three miles from here. He threw his bundles down and made for the nearest tree, which happened to be a large persimmon tree loaded with ripe fruit, into which he quickly climbed, knocking off some of the ripe persimmons as he made his way up into the tree. When the snake arrived at the foot of the tree instead of climbing up after the negro, he began to eat the fruit that had fallen to the ground. After the snake had finished devouring all the persimmons in sight he began to coil himself around the tree preparatory to climbing it. The frightened negro, seeing the snake had eaten the fallen persimmons, gathered a handful and threw them to the ground; then the snake uncoiled from around the tree and began to eat the fruit the [man] had thrown down.

This was repeated several times, when the [man] bethought himself of a small vial of morphine which he was carrying home to his sick wife. He split open several persimmons and emptied the contents of the vial upon them and dropped them to his snakeship, who immediately ate them. The drug acted like a charm upon the snake and he was soon lying helpless upon the ground. The negro climbed down and ran to a house near by and secured help and returned and dispatched the snake lying helpless from the effects of the morphine. The snake was eighteen feet long, six inches in diameter and had forty-four rattles and a button on the end of his tail, and had been the terror of Tubb's Creek swamp for the past twenty-five or thirty years.—Zainesville (Ala.) *Messenger*.

Indiana, Pennsylvania, *Democrat*
November 29, 1888

Monster Snake Bends Iron Bar

Holding forth on his wooded throne along the banks of the lazy Wabash surrounded by a royal retinue, the largest rattlesnake in the world no doubt is living, for the farmers in the locality of Grayville, Ill., recently have been served a taste of the monster reptile's strength.

The monster rattler recently made an incursion on the Brey farm and there devastated property for several hours. When discovered by Brey the reptile was plowing up rows of potatoes for the frightened farmer who flung an iron bar at the reptile. The bar weighed fifty pounds.

Although Brey hit the snake, it curled itself around the bar and carried it off, taking such hurdles as hedges and rail fences in its flight. More than a dozen rails and three yards of hedge were destroyed. At last the reptile released the bar which was bent into a horseshoe shape.

Brey, near whose farm the retreat of this snake is located, has for more than three years worn rubber boots to protect himself against snake bites. The rattler's hundreds of scouts are constantly on the watch, and thousands of snakes are said to abound along the river banks

Dynamited, fired upon, his retreat subjected to every known method of known extirpation, the great snake defiant of them all, has withstood the attacks with seeming impunity. Hundreds of small rattlers have been killed this summer by the angry farmers of the neighborhood, but never have they even injured the terror of the Wabash.

Washington, D. C., *Post*
August 26, 1907

The Lineville *Tribune* can take the whole bakery. Little story tellers must now take a back seat, for the *Tribune* is the champion. Here is what it told last week: "A wild and weird, but o'er true snake story comes to us from Marion, and is vouched for by an eminent and truthful citizen of that burg, and runs as follows: On September 13th Joe Gann, while loading hoops in the timber near that place discovered and succeeded in capturing a hoop snake 21 feet 4 1/2 inches in length and 1 foot 9 1/2 inches in diameter. When first discovered his snakeship attempted to get away from his pursuer in the usual method of hoop snake—by throwing himself into the form of a hoop and taking his tail in his mouth rolled down through the thick timber in the direction of Weldon with great velocity; but being hotly pursued he evidently made a "miss-cue" with his steering apparatus and his tail striking a large white-oak tree about 10 feet from the ground squarely in the center drove the large, sharp and polished horn with which his tail was armed through the head of the snake, and into the solid tree with such force as to split the body of the tree for several feet and impaled the reptile so securely that his capture was an easy matter. The horn which our informant sent along to this office by Mr. W. P. Sullivan, of this place, and which ornaments our sanctum, was somewhat shattered by the force of the blow in striking the tree, but is otherwise intact and much resembles in size and shape the horn of a two-year-old heifer. Incredulous persons who may be inclined to doubt the truth of this story can call at the office and see the horn for themselves."

Humestown, Iowa, *New Era*
September 25, 1889

Fortunate Escape

On Friday evening last, a party of young men from this place were ascending North-Newport river, in a small boat, and near the head of navigation, discovered an enormous SNAKE making towards them. Four muskets, three rifles and a pair of pistols were discharged in quick succession at the monster, which arrested his progress until the party re-loaded and fired a second time, this, with the aid of harpoons, bayonets, &c. succeeded in taking life. The Snake measured in length 21 feet and a half, and 18 inches between the eyes. On opening the body, a negro's head, a calf, four alligators (each measuring three feet) a green turtle, two dogs, six geese, besides many small birds, were found therein. A similar animal has lately been seen in Sunbury river, by some of the young men, while on their nocturnal rambles. From the description heretofore given of the great *Sea Serpent*, and the enormous size of this animal, many have no doubt in saying it is the same. Messrs. Editor, by giving the above a place in your paper, you will confer a favor on many Subscribers.

Riceborough, (*Gaz.*) May 29, 1820

Gettysburg, Pennsylvania, *Adams Centinel*
July 12, 1820

A Texan Snake Story.

Bud Brown who resides a short distance north of this place, had an exciting experience a few evenings ago with a huge serpent. Bud, it appears, was on his way home along the Missouri, Kansas & Texas track, from this burg, and had just reached a point about one and a half miles north of Holland, when he heard a queer bellowing sound similar to that made by alligators in the dead of night. He looked behind him, and, shure enough, saw a sight that for the time being rooted him to the earth.

About 100 yards in the rear he could plainly discern a snake of monstrous size. The reptile approached him at a rapid rate, coming so close to him, in fact, that Bud, with a shriek of terror, jumped from a high trestle, landing in the soft sand below, escaping, fortunately, with only a sprained ankle. Luckily for Bud, some cattle were grazing along the track, which the big snake at once attacked, making a way in a jiffy with a calf and disappearing in the timber.

Bud, who is a worthy citizen and temperate withal, estimates the length of the snake at 42 feet, with a body as thick as a telegraph pole. He also avers that the monster was adorned with two formidable fangs, and, in addition, was equipped with horns about a foot and a half in length.

Recently farmers in this vicinity have missed calves, sheep, and pigs, and it is now thought, and with some reason, too, that they have been gobbled by the horned monster.

A posse of brave young men, well armed, has been organized, and they will scour the woods until they make mince-meat of the thieving big snake.—Holland (Tex.) Corr. St. Louis *Globe-Democrat.*

Humeston, Iowa, *New Era*
August 13, 1890

A Terrific Snake Story.—Mr. Joshua Buddington, of North Attleboro, has furnished the Providence (RI) Chronicle, with the particulars of a snake capture on his farm on the 10th inst. A cow had been missing for several days, and his son, while searching for her, found her laying dead, with an enormous snake entwined around her hind legs. The lad ran home, and a party sallied out to slay the monster. They found that the snake had made an incision into the left side of the udder, through which he had inserted his head about four inches, and was in the act of extracting the milk at the time the party arrived. He was immediately despatched by a tremendous blow from a club given in the region of the neck. On being struck he at once uncoiled from the limbs of the cow, drew forth his head, and after gasping three or four times, expired.

His back is zebra striped, and the belly of a dark green, with small spots thickly interspersed. Around the neck and directly back of the jaws, are four stripes or rings of a bright yellow color, and just under the throat a small bag or hollow membrane is suspended, filled with a thin liquid substance. This membrane is perfectly transparent and through it the appearance of the contents is dark green. The length of the snake is 14 feet 3 inches—circumference around the largest part of the body, 1 feet 10 1/2 inches, from the end of the upper jaw to the eye, five inches—width of the head, which is very flat, 7 1/4 inches. The species to which the snake belongs has not been ascertained.

Gettysburg, Pennsylvania, *Adams Sentinel*
July 31, 1843

Snake Charming.

For the last two weeks a son of Allen Rogers, aged eleven years, a woodcutter on the Blue Mountains, about three miles from Hamburg, has been in the habit of leaving is father's house every morning about 9 clock, and not returning till noon. The parents of the boy have questioned him several times as to where he went, and the boy would reply, to play with a neighboring boy named Springer. On Friday last the father watched his son, and followed at a short distance, and when about a half-mile from the house, the boy entered a piece of thick sprout land, in from the road some two hundred yards, where he seated himself upon a large rock, and in less than ten minutes the father was horrified on seeing a monster black snake crawl upon the rock and put its head on the boy's lap. The father states that the snake was the largest he ever saw on the hills. He states that it was easily fifteen feet long and as thick as his arm, which is well developed. The boy had taken bread with him and was feeding the snake, which at intervals would stick a large tongue out as if hissing for more to eat. Then it would coil itself around the neck and body of the boy, and play with its mouth and neck with the boy's hand. The father had often heard of snakes charming children, and that if they were disturbed while they were in the act, they would kill the child. As the father turned to leave his boy with his deadly companion, he turned back, and the snake hearing a noise, at once uncoiled itself and raised its body at least four feet from the rock and looked in all directions, and then it returned to the boy's lap, and the father returned, home and waited the boy's return, which was, as usual, at noon. When told that he had been playing with the snake, the boy said the first morning he met the snake he liked to play with it; then he took it food, and he was so much pleased with his companion that something told him that he must meet the snake every morning. One morning he said he was late, and when he reached the place the snake was standing up, and it came out to meet him, then followed him to the rock. There is something very strange about a snake charming not only children, but I have heard of adults coming under their charms. There is certainly some truth in the fascinating powers of snakes.

On Saturday morning the father and two of his neighbors went to the place with guns, and at the usual time the snake made its appearance when all fired at one time, killing the charmer.—*Reading Eagle.*

Hagerstown, Maryland, *Mail*
October 29, 1875

Fight With A Snake.

C. B. Stevenson, of Pittsburg, who is at present in the oil fields at Knoxville, has them all flagged when it comes to a genuine hair-raising story. Stevenson weighs about 250 pounds and is not used to strenuous life, being possessed of some money and ill trained for the mortal combat in which he accidentally became engaged.

While wandering around over the oil territory, Stevenson came across a romantic spot, cool and shady, so inviting to his tired and corpulent body that he desired to rest awhile. While he was thus reposing on the log he felt something touch his back, but paid little heed, thinking it but a stray dog. Finally he turned around and saw the largest monster in the snake line he had ever beheld in or out of captivity. The demon seemed to laugh at him from its wicked, fiery eyes, as it edged toward him, while he seemed rooted to the spot. Gathering his nerves by one supreme effort, he jumped over the reptile and emitted a yell that would have caused a Comanche to turn livid with envy, while he sped down the hill in huge, reckless lunges that were extremely trying on the seams of his garments.

A glance back showed the snake to be gaining, and realizing he could not keep up the killing pace Stevenson took to a young sapling and shinned up, only to be followed by the snake. He dropped to the ground and grabbed a club and decided to die game. The snake reared at him and the battle was on. At first he could not hit the elusive and wiry demon. He finally got in a crushing smite on the animal's back and the fight was over, while he dropped from exhaustion. Those who saw the snake are at a loss to classify it. One native writes to a friend in this city that it was a nine-foot blacksnake with horns and ears and that it had thirteen buttons on its tail, with pink and green stripes along the back.

Indiana, Pennsylvania, *Messenger*
September 16, 1903

Rainforth's Missing Child

Cor. New York *Sun.*

West Union, Adams Co., Ohio, is near Ginger Kidge, a rugged, sterile upland, which is much excited over the killing of an enormous black snake which for several years has played havoc with the farmers' flocks. Hogs, poultry, calves, sheep, etc, have mysteriously disappeared, always at night. Two years ago a band of gypsies were camped in the neighborhood and they were accused of stealing the missing property. John Rainforth, a farmer who greatly suffered from these depredations, swore out a warrant before 'Squire Peter Anns and had several of them arrested. They had a preliminary examination, but nothing was proved against them and they were discharged. They went away muttering threats of vengeance.

Mr. Rainforth had a golden-headed little daughter, four years old, whose beauty and sunny temper were the pride of her parents. On the day after the arrest of the gypsies little Nellie Rainforth was missed. She was last seen playing on the edge of a rocky ridge a short distance from the house. Search was made for her, but neither she nor the lamb was found. The whole neighborhood was aroused and men scouted the fields and woods for miles around. Mr. Rainforth suspected the gypsies of abducting her out of revenge for their arrest and followed the party across the Ohio river into Lewis county, Ky. When he came up with them they indignantly denied all knowledge of the child's whereabouts and a search of their camp failed to discover his little daughter. He returned to his home broken-hearted.

One day last week Mr. Rainforth was planting a field of about twenty-five acres situated near his house. He had not been at work long when he discovered what at first seemed to be a fresh furrow across the middle of the field. He stopped work and followed the track to a fence which separated the field from a dense thicket of underbrush. On the fence he found blood and some sheep's wool which at once convinced him that the body of a sheep had been dragged across the fence. He went to his pasture and found that a large Cotswold ram was missing. Accompanied by four or five neighbors, Mr. Rainforth made search for the missing sheep. The track through the brush was marked by drops of blood and tufts of wool. About sixty rods from the fence they came to a ledge of rocks forming one side of a steep hill. The track led directly to this ledge in which was found an opening of sufficient size to admit the body of a large man. A large charge of giant powder was exploded in the opening and the rocks were thrown asunder by the blast. When the smoke cleared away the farmers drew near and peered down the opening, and there, among at least a wagon load of bones, lay a huge black snake, quivering from his hurt. The farmers waited until the snake was dead, and then attached a chain to his body and dragged the monster out of the hole. He measured fifteen feet seven inches in length, and the biggest part of his body was over two feet in circumference. He had an ugly looking head and enormous fangs, sharp as needles. The missing ram lay beside him, crushed out of shape, and covered with a sticky, glutinous substance.

I visited the spot to-day and saw the monster snake. While I was there men were at work clearing the den of the bones. In a corner one of them picked up a human skull. It was small, like a child's, and he brought it forward to the light. Mr. Rainforth was standing by my side when the man came toward us with the skull in his hand. He glanced at it, and, staggering against a tree, buried his face in his hands and burst into tears.

"Poor little Nellie," he cried, through his sobs. "My God, it is horrible!"

After a time he controlled his feelings and told the story of his daughter's mysterious disappearance two years ago. The bones of the little one were gathered together and buried in the family plat in the cemetery at West Union. The discovery was kept from Mrs. Rainforth, for the poor woman has never ceased to mourn for her lost child, and her husband feared that this intelligence might seriously affect her, she being in delicate health. There can be no doubt as to the identity of the skeleton, for a gold chain which she wore around her neck was found among the bleaching bones.

Cambridge, Ohio, *Jeffersonian*
July 15, 1880

A Veritable Snake Story.

An officer of the army, who served with distinction in the Florida war, was and still is in the habit of delighting the mess table with his reminiscences of scenes which occurred there. On a recent occasion, when snakes, alligators, and other objects of the reptile genus, which flourish so extensively in that garden of the world, became the subject of conversation, he related a circumstance apropos to snakes, which happened to himself. One day, said he, I shouldered my gun, and went in pursuit of game. In passing through a swamp I saw something a few feet ahead of me, lying on the ground, which had every appearance of a log, it being about forty feet in length, and about a foot in diameter. So positive was I that it was nothing but a log, that I paid no attention to it; the fact is I would have sworn before a court of Justice that it was a log, and nothing else. You see I had never heard of snakes growing to such a huge dimension, and the fact is, I never should have believed it if I had. Well, he continued, between me and the log, as I took it to be, there was a miry place, which it was necessary for me to avoid. I therefore placed the butt of my gun on the ground right ahead of me, and sprang upon it, and lit right on the top of—what do you suppose?
"A boa constrictor," said one.
"No."
"An Anaconda."
"No."
"What could it have been?" enquired a third.
"Just what I supposed it to be—a log," said the wag.

Gettysburg, Pennsylvania, *Republican Compiler*
April 5, 1858

Steuben *Republican*: "Some one of this place has been stuffing the editor of the Fort Wayne *Dispatch* with a snake story describing the capture of an enormous reptile, of a spotted color, measuring twenty-one feet six inches in length. The story is a pure fabrication from beginning to end."

Fort Wayne, Indiana, *Daily Sentinel*
May 9, 1883

Rumors

No Truth in the Circleville Snake Story.

That monstrous story of a monster snake being seen near Circleville, this county, recently, is still on the wing, growing like an avalanche. The writer of these lines saw Mr. David Ayers, a few days ago, and asked him about it. All he knew is what he has seen in the papers. Yet it is reported that it was on his farm the reptile has its home, and it was to him the veracious boys made report, according to the story, of seeing the snake crossing the highway.

Of course there is no truth whatever in the story. There is even no foundation for it, Mr. Ayres says, except that about three years ago he saw a large blacksnake on his farm. He hasn't seen it since, nor heard of any who has.— *Union.*

> Middletown, New York, *Daily Press*
> July 28, 1893

Babies Fed to Snake

The United States grand jury at Albuquerque, N. M., is investigating reports that the Indians of the Pueblo of Zae, the most isolated of the Pueblo tribes of New Mexico, feed a certain number of newborn babies each year to a mammoth snake which is worshiped by the tribe. The interior department will probably be asked to interfere.

> Middletown, New York, *Daily Press*
> December 21, 1905

No surprise, some rumors lend themselves to commercial application...

Chased by Monster Snake

Two well known Syracusans just returned from the Thousand Islands tell of being chased by a monster sea serpent while fishing at Eel Bay. They refused to name the brand of liquid bait used,

but acknowledge that they had a box of their favorite smokes—
the new Full Dress clubhouse shape, Havana wrapper cigars—
nickel goods, but worth a dime. Order box for your vacation off
any dealer.—*Adv.*

Syracuse, New York, *Herald*
July 4, 1911

Old Deed Cites Paul Bunyan of Snakedom

Mansfield, Ohio.—(U.P.)—WPA workers engaged in a survey
of records of Richland county discovered an 80-year-old deed
conveying not only a 20-acre tract of swamp land but also a fabulous giant rattlesnake believed to dwell in the marsh.

By deed dated Feb. 10, 1858, Geo. B. Wright, as receiver for a
railroad company, conveyed the land and the monstrous snake
to Allen B. Beverstock, of Lexington, Ky.

The reptile was supposed to be of such size and strength that
it could push down fences, breaking the rails with its weight.
Beverstock not only had a clause of conveyance inserted in the
deed to gain title to the serpent, but pen sketch of the animal was
drawn on the face of the deed.

The picture was more than 10 inches long. It was that of a
gray snake, its back splotched with yellow, with a longitudinal
row of black spots bordered with white. What became of the snake,
or whether it ever was seen, was not recorded.

Hammond, Indiana, *Times*
February 15, 1938

Showman's Hype

Mammoth Snake Story
Missouri Has Largest Snake in U. S.—
Story Vouched For By Oxnarders

Below we give the story of an immense snake which was
captured in Polk County, Missouri, some two or three years ago.

While the story of the mammoth reptile seems a little large, it is vouched for by several men who came from there. There are several Polk County men in the city who claim they knew about it, but it was not thought much of a curiosity there. Frank Inglis is one of these, and he avers that snakes nearly as large as this one were a common sight there. Judge Short, who lived just across the river in Illinois, claims he knew of the snake well and at one time saw it while on a trip to Missouri. That was some ten years ago, however, and it was not as long then by some ten feet. Milt Nicholson used to go down into Missouri courting the girls when he lived in Iowa and often heard of this particular snake, and one night as he was going home late he came across it, but it got out of his way before he could kill it. Ed M. Sheridan, the Ventura novelist, avers that he saw the snake once when he was visiting his brother in Missouri, but he laid it to the beautiful "moonshine" which they have there. Tom Rice says he left Missouri many years ago, but at that time it was over four hundred feet long and used to come to his father's pasture and swallow calves whole. The snake will be on exhibition at the Portland exposition this summer, so that all doubting Thomases can see for themselves. The name of the writer of the article is withheld because of his extreme modesty:

The story dates back to 1886, when a monster snake infested Polk County, Mo. My father lived several miles from Brighton, in that county, and an old cousin of his was a member of our family. One day as the cousin was on his way to a horse race, he encountered an obstacle at a cross roads which puzzled him. Something was lying across the road, which he first mistook for the trunk of a tree which had been uprooted during a storm that had prevailed the night before. But as he stepped upon it to surmount it the thing moved, and he quickly drew back, convinced it was a snake, although he could see neither its head nor tail, its length extending so far each way into the dense brush that he could not see where it terminated. While the old man was impressed with the marvel he had seen, he was a true sportsman and much more interested in the races than all the serpents in the world, so he resumed his journey to Brighton. There he told of his adventure, but his narrative was received with incredulous shrugs, and some persons were harsh enough to suggest that he had been drinking, and that the snakes he saw were in his boots. However, he stood his ground, and ventured the opinion that time would vindicate him. And it did.

Hogs, sheep, calves, cows and horses began to mysteriously disappear in that section of Polk County, and it was evident their vanishing was not due to the industry of thieves, of whom there were numbers in that vicinity at that time. Where they went was the question, and it was not answered until Bill Barham found (and badly scared he was) the solution of the riddle. He was driving a span of horses on the road which my father's cousin had walked that eventful morning and was suddenly confronted by what looked like a cave with a moving roof, but which in truth was the gaping mouth of a huge serpent, which quickly engulfed the horses and wagon, the driver owing his escape soley to the agility he displayed in leaping from his seat when he saw the roof of the "cave" closing upon his outfit. He ran a short distance, then fainted, and when found several hours later and revived could hardly compose himself sufficiently to recount his extraordinary adventure. There was no longer any doubt of the existence of a mammoth snake; its huge trail through the woods could be plainly seen, and the severe losses sustained by owners of live stock were eloquent in testimony of the voracity of the monster. Searching parties were organized, but they could not come upon the snake, until in October last year some cowboys took up the matter and led by Oscar Mitchell, with whom I am well acquainted, decided to capture the snake if it took all year.

Impressed with the evident size of the snake, they laid plans accordingly. It was found that the serpent kept quite close to the one trail, which extended from southeast to northwest. It was shown that the animals which disappeared had always crossed this trail. Utilizing this knowledge, the cowboys procured inch cables of very flexible, but exceedingly strong rope, and placed three large nooses at intervals in the path of the serpent. Then they constructed an equal number of capstans with which to wind up the cable, and stationed watchers at each one, with horses hitched to the mechanism, ready to wind up at a moment's warning. An old horse was placed on the trail as bait and the watch began. It proved to be a weary one. Night after night passed, but no snake appeared. There was, however, no relaxation of vigilance and the crews and horses at the capstans were relieved at short intervals, there being no hour in the day when the same were not fully manned. Finally patience had its reward, and the snake came. It seized the shackled horse, and while it was devouring it, the word wan given and the capstans were revolved. There was a tremendous wiggling and threshing, trees were uprooted and the earth made to heave like the billows of the ocean, but the serpent had crawled through all the nooses, and it was finally strangled to death, "before it know," as one of the cowboys expressed himself.

Accurate measurements showed it to be 842 feet long, and of corresponding girth.

The next question was what to do with the monster, and it was finally decided to prepare it for exhibition at the Lewis and Clark exposition. First its oil was extracted, and there was enough of it to last the people of Polk County for centuries. A shed has been built over the carcass, and it will be converted into a sort of serpent passenger train. The marrow was extracted from the vertebrae of the backbone, and the hole thus caused is large enough for a man to crawl through. Strong cables will be passed through the vertebrae and windows will be pierced through the skin, which is magnificently and gorgeously colored. The great question is, how to get the exhibit to Portland, but the solution is practically assured. A railroad spur will be built from Wishart, the nearest railroad point, to the place where the snake's body lies. The prepared carcass will be lifted onto specially constructed trucks and transported to Wishart, whence the journey will proceed over the main lines of railroad via St. Louis, Denver and Salt Lake City. It is expected that the snake will be ready for transportation by next Independence Day, and it is the general sentiment in Polk County that Gov. Folk could hardly do a more graceful thing than with his staff, attend the celebration which will be held that day at Brighton, and thus make more memorable the start for the exposition of this—the greatest natural curiosity the world can show.

Oxnard, California, *Courier*
February 10, 1905

Social Control

A Snake Story

The Barnesville *Enterprise* is entitled to the "champion belt" for the biggest snake story. the following is copied from the last number of that paper.

Another Monster—A Man Takes a Ride Upon a Snake—A few days since, while Joseph Selby was gathering raspberries, he came to what he supposed to be a log, and being somewhat tired he sat down upon it to take a rest, when to his surprise, he commenced moving down the hill. He was so much frightened that he did not know for some time what was the propelling power, but when he recovered himself he found that it was a monster snake that was carrying him upon its back. He supposed it was from fifty to sixty feet long and as thick as his body. He fell off during the journey, and the snake continued down into the hollow. This snake has been seen by various parties for several years past, mostly in the raspberry season, but nobody has enjoyed such an intimate acquaintance with the monster as Mr. Selby. We would advise everybody to be on the lookout for this reptilian monster during raspberry time.

Coshocton, Ohio, *Democrat*
August 12, 1873

Fiction

Hoffman's "Snaik."
The "Gazette's" World Renowned Explorer Again Exhibits
His Courage and Kills the Monster Serpent that Has Lately
Bothered the People of Elkhart County.
He Chops Off the "Snaik's" Head After an Hour's Steady Work—
A Cowardly Attempt to Rob the Truthful Hoffman of the
Glory of His Achievement.

Roann, Ind., August 19, 1885.
To the editor of the *Gazette*.
A few days ago we got word that we were wanted in Elkhart county; that a mammoth, ill-shaped serpent had been seen by

some of the sturdy yeomanry of the territory that comprises that country of the north. We hastened to prepare our "grip" and say goodbye to wife and babies, and in less than no time we were on the popular Cincinnati, Wabash & Michigan, going with lightning speed to what is called the county seat of a very fine county.

After a moment's rest we sought his honor, the mayor of the village. He was very glad to see us for he was having grave fears of the monster snake that was and had been playing havoc with crops and stock in the southwestern part of the county. We asked him if he knew of the whereabouts of his snakeship at that time. His honor said that the last seen of him was plowing his way down the roaring St. Joe, but he had business that day and could not go with us. We asked a number of the dignitaries of the village to lend us what aid they could, but all seemed to have something to do. We saw plainly that these Elkhart folks were very cowardly. At this stage of the matter we resolved to go to work alone and see what could be done.

We started down the south side of the river, into the neighborhood where his snakeship had been seen. We soon found the sturdy old granger that had seen and gave such a vivid description of the monster. His hair had turned white as snow in less than a fortnight and deep furrows were spoiling a round chubby face that had not seen more than forty summers. We interviewed him about the object of our search. At the mention of the snake he turned deathly pale and shook as if he had palsy, but partially recovered when we told him who we were and that we had come to rid the country of the monster. He quietly informed us that the general supposition [was] that the serpent had gone to Marshall county and was at that time basking in the marshes of that territory. He seemed quite certain that such was a fact, as he had heard of a great tumult near Plymouth, the county seat of that county of the bog.

We asked him to take us and our grip to that town. The poor granger tremblingly told us that he could not think of such a thing. He was quite sure if he ever did get sight of the snake again he would be a dead man. Our pity for him admonished us not to press the matter upon him. We then trudged along on foot toward the town of marsh fame, feeling quite sure that we would find the object of our search somewhere along the raging Yellow river. Late in the night, foot-sore, hungry and tired, we saw through the low brambles and high grass a few lights only a few rods ahead of us. This was Plymouth.

As we entered the long and zig zag street or road that runs through the town, we heard a loud, screeching noise that sounded different from anything we had ever before listened to. As we stopped at the only hotel, which is a fine hewed stone structure, we hastily inquired as to the noise just a few rods from the main road upon which the town was built. We were told that the wild geese, ducks and cranes, and all the water fowls in the state, had summered there, and that in the last few days a large snake from Elkhart county had passed through the pond, and had created an unusual commotion. We said no more then, but were elated over our success so far in hunting the monster down. We were soon fast asleep on a good bed in a room whose ceiling could not have been over four feet high. In the morning we sought the marshal of the city of swamps, and asked him what he knew about the monster snake that had passed through the duck pond. He not being a man of vast intelligence, our gain of knowledge was rather limited; yet we learned enough to be certain that the Elkhart monster had gone in the direction of Rochester.

This was sad news to us, for the general lay and make-up of the land that composes Fulton county is such that the very devil himself could hide there for a thousand years to come, but we again took the road for the village on the bank of Devil's lake, We were not long in reaching that benighted town, but to our surprise the place was as still as the grave. Windows and doors wore closed and not a living soul could be soon anywhere. We stood in the center of the place for full one hour, thinking that at last some one would make a stir, but not so.

We now began too look around and wonder what to do next. In looking down a short, crooked alley we saw the word "Sentinel" printed on a clapboard and nailed on a small log house just at the end of the crooked alley. We approached the door, which was very low and fastened with a pin, and gave a loud knock. For a few minutes all within seemed silent as the grave. Just then we heard scrambling on the straw roof, and on looking up we saw a woolly headed and dirty faced man peeping over the comb of the roof at us. His face wore savage grin. He asked who we were and what we wanted. We answered his questions as hastily as they were asked. He seemed glad that we had come, and told us that his name was Tully Bitters, and that he was editor of the Rochester *Sentinel*, and this was his office. The reason his office was closed was that the large snake seen in Elkhart county had come down to Rochester and scared everybody nearly to death; that he had not been out of his office for three days, and that he had about lost his mind, fearing that the monster would take the town. He seemed overjoyed when we told him that we were on hunt of the monster, and that we were bound to take him, either dead or alive, before we gave it up.

Mr. Bitters kindly offered to assist us all he could, provided we would not ask him out of the office. We assured him that we must have some help, and that if we did not get it, we would abandon the chase. He asked us to go and see his brother, who also was an editor, and see what he thought about it. This we flatly refused to do, and with some warmth said if he was too much of a coward to lend us some assistance, that we hoped that the serpent would eat him up, but we were assured at the same time that no snake would tackle him in his present plight, for he looked decidedly tough. The hair had slipped from his head, and his skin had turned a dirty green on account of the malarial which settles all over that country. He resembled a digger Indian very much.

The men resolved to go down to the lake and see if we could find any trace of the object of our search. We did not go far until we met a farmer riding a horse at the top of his speed, yelling, "A snake! a snake!" We hailed him and inquired what was the matter. He told us that a monster snake had taken possession of his premises and had actually eaten two horses and three cows for him in the last hour. We did not wait to hear any more, but told him to go to the Sentinel office and tell Bitters to come at once. The farmer went his way on a fast run and we repaired to the farm. Here we found the snake in just such a condition as we expected. He had overloaded his stomach, and was lying out in a large field perfectly helpless. We walked all around him and examined him very closely. He was a very large snake. We sat down under a shade tree near by to consider, and pretty soon the farmer came back, but was entirely alone. He said that he could not find a human being in the whole town, and wondered what was to be done. We told him to got an axe and we would show him what to do. He got the axe and we took it and proceeded to out the snake's head off, which we did after one hour's hard chopping, the axe not being very sharp.

By this time it became noised around that we had both seen and killed the monster, and for the time we wore considered the heroes of the hour, until Bitters came within half a mile of the spot and sent a boy to tell us that we were stealing all his glory from him. Had it not been for him coming out on the straw roof of the *Sentinel* office and telling us what he knew about the matter, we would have never found that snake and that he claimed all the glory himself and wanted it distinctly understood that he was the hero, not only of the hour, but of the whole county. We did not dispute with the poor silly follow, but picked up our grip and meandered toward home, leaving him in all his glory.

A. W. Hoffman.

Fort Wayne, Indiana, *Weekly Gazette*
August 27, 1885

Paranormalism

Paranormalism is the tendency to ascribe mysterious—even super-natural—qualities to natural events. Even if an investigator assumes that an unknown species of snake is a possible explanation for certain reports, that is not outside scientific possibility (even if it goes against the opinion of the majority of biologists), and so shouldn't be relegated to the paranormal. Most discussions of giant snake stories in cryptozoology have focused on natural explanations, but a few cases have wandered off track. One report, for example, popularized by John Keel, has been touted as mysterious due to an unusual injury caused by the encounter.

A 32-year old Forest, Ohio, man claimed he was riding horseback when an eight-foot long black snake "struck at him, then wrapped its body around his leg, crushing his ankle." (Lima, Ohio, *News*, June 11, 1946.) More likely, the ankle was broken when he fell from the horse. He was treated at a Kenton, Ohio, hospital. Keel (1994) stated that the man had to have part of the foot removed surgically and was confined to his bed with feverish symptoms for a year. So, of course, the story then continued to be reprinted by other writers with vague allusions to high strangeness and whatnot. Now, anyone who has a background in biology, anatomy, or medicine should be able to point out that trauma to the human body can create ongoing complications. There's nothing remarkable about that. Ankle breaks can result in small bone fragments that have to be removed, and one has to wonder about the rehabilitation therapy of the 1940s—even today it can be months before healing is complete. There was also the possibility of infection, which could certainly create further medical problems.

While I am not at all convinced that a snakebite even occurred in this instance (I doubt that an 8-foot colubrid has the teeth to break through a boot), there are parallels to a case from the former SFRY (Yugoslavia). Maretic and Russell (1979) reported that a 29-year old man stepped on a large snake while walking along a road. He was bitten, and frightened, ran up a hill. "As he ran, he noted his right leg was becoming 'more heavy'. The pain increased in intensity and began to radiate to his right knee and thigh. He stopped and in agony vomited twice." After hospitalization and a battery of tests, it was finally determined that upon being bitten, the man must have jumped, which caused him to fracture a bone. Running up the hill increased the trauma and pain. It is likely he was only bitten by a common non-venomous species, the Aesculapian snake.

While not strictly paranormalism, there is another error of misinterpretation that creeps into certain writers' speculations—the insistence on forcing modern perspectives on historical cultures. Often in cryptozoology, researchers point to Native American folklore and artifacts that may offer early cultural recognition of unknown species. To an extent, there are valid points to be made, but there are also a number of cases where the evidence pointing in another direction is far stronger. When speculating on whether "big owls" may be referenced by Native American legends of the Flying Heads, for example, it is just plain foolish to ignore their own interpretations of these mythical creatures as tornados.

When discussing giant snakes, it's no surprise that some of the large serpent effigy mounds and other rock art are sometimes referenced. I grew up in southern Ohio, with close ties to the area near Adams County and the Great Serpent Mound. It is frustrating to see the ignorance of archaeological research that is too often displayed as some researchers entangle these treasures in hodge-podge theories about giant snakes. Suffice it to say, in most cases the impetus behind these serpentine structures lies in the stars—the Great Serpent Mound and numerous other effigy mounds are strongly ethnoastronomical.

Known Species

North America (north of Mexico) currently has 141 recognized native snakes (CNAH 1994-2007), incredibly diverse in morphology and behavior, though this number is bound to change as herpetologists investigate species and genera more closely. We've recently seen several groups undergo revision, particularly the ratsnakes. The United States also has three exotic species with breeding populations, two of which are significant in size (*Boa constrictor* and *Python molurus*, both in Florida).

Of our native species, the recognized maximum lengths in colubrid snakes are (Boundy 1995; Devitt, et al, 2007):

Florida Green Water Snake
| *Nerodia floridana* | 1880 mm | 6 ft 2 in |

Racer
| *Coluber constrictor* | 1905 mm | 6 ft 3 in |

Mud Snake
| *Farancia abacura* | 2070 mm | 6 ft 9.5 in |

Common Kingsnake
| *Lampropeltis getula* | 2083 mm | 6 ft 10 in |

Western Ratsnake
| *Pantherophis obsoleta* | 2184 mm | 7 ft 2 in |

Pine Snake
| *Pituophis melanoleucus* | 2286 mm | 7 ft 6 in |

Eastern Ratsnake
| *Pantherophis alleghaniensis* | 2565 mm | 8 ft 5 in |

Coachwhip
| *Masticophis flagellum* | ~2591 mm | 8 ft 6 in |

Indigo Snake
  Drymarchon corais   2629 mm   8 ft 7.5 in
Bullsnake
  Pituophis catenifer   2667 mm   8 ft 9 in.

For venomous species, the recognized maximum lengths are:

Cottonmouth
  Agkistrodon piscivorus   1892 mm   6 ft 2.5 in
Timber/Canebrake Rattlesnake
  Crotalus horridus   1892 mm   6 ft 2.5 in
Western Diamondback Rattlesnake
  Crotalus atrox   2337 mm   7 ft 8 in
Eastern Diamondback Rattlesnake
  Crotalus adamanteus   2515 mm   8 ft 3 in

So how might one of our larger snake species achieve greater lengths? A snake's length is determined by various factors. Individually, large snakes may result from a plentiful food source (particularly during early growth periods), consistently larger prey size, or (rarely) a genetic quirk. A population of snakes may exhibit gigantism due to genetic isolation, environmental factors, and/or larger prey size. After studying a recently isolated population of Australian tiger snakes that showed a significant increase in body size, Aubret and Shine (2007) suggested that "geographical divergence in mean adult body sizes in this system initially is driven by a rapid shift due to phenotypic plasticity, with the divergence later canalized by a gradual accumulation of genetic differentiation."

The availability of the correct-sized prey is particularly influential. Young snakes generally allocate energy resources to growth, because the larger a snake is, the larger the prey it can take (Madsen and Shine 2002). So, the more abundant the (easily-captured) prey, the faster the snake will grow. With abundant prey, the limiting factor for snake growth then swings to the size of the available prey. When the choice is offered, many snakes shift their feeding preferences to a larger prey species as they grow. In several species, large snakes drop small prey altogether, probably because smaller prey can be difficult to catch and manipulate as the snake grows larger (Keogh, et al, 2005). (Conversely, some species, like ratsnakes, exhibit "ontogenetic telescoping," where their opportunistic and generalized feeding habits merely expand to a wider range of prey (Weatherhead and Blouin-Demars 2003).)

Dr. Henry Fitch (2006) noted, "The large size of the Bullsnake is likely an adaptation for it to prey on the relatively large pocket gopher. ... A promising subject for studies still to be made is the correlation of snake size with that of each of the kinds of gophers with which [the *Pituophis* species or subspecies] is sympatric." Madsen and Shine (2000) noted that early rapid growth due to abundant prey has a "silver spoon" effect on some snakes, giving them not only an initial advantage, but having a long-term effect on their growth rates and maximum adult body sizes.

As they reach maturity, snakes then allocate energy to reproduction rather than just to growth. Because snakes exhibit asymptotic growth (Mori and Hasegawa 2002), or continued growth after maturity, just how much they grow again depends on diverse influencing factors. Here, too, predatory tactics are important, as dependence upon smaller prey species limits a snake's ability to maintain the energetic costs of both reproduction and growth. If the snake can hunt abundant larger prey, for example, it has a better chance for continued growth. This pattern isn't limited to snakes, of course. Jessop (et al, 2006) found that Komodo dragon populations with large prey locally abundant had significantly larger maximum lengths than those with low prey density.

Genetic isolation in the contiguous United States and Canada is uncommon in snakes, but there is at least one case where a wide ranging genus can be found in small disjunct populations (e.g., *Pituophis*). When reviewing body size trends accompanying isolation on islands, Boback (2003) noted that the majority of vipers, and half the boids and colubrids in the study exhibited dwarfism. Eight out of eleven species that exhibited gigantism were colubrids that gorged seasonally on seabird chicks. So, it is unlikely that isolation itself is a good predictor for gigantism in North American snakes. But when a rapid size increase occurs, it can happen quickly with minimal changes in the genotype. In Aubret and Shine's (2007) study, gigantism occurred within 100 years, while Keogh, et al, (2005) noted that the genetic divergence between a population of normal-sized mainland tiger snakes and nearby island dwarfs and island giants was only 0.38%.

What influence would latitude have, given that (as will be seen in the genus/species maps) many of the larger snakes have large ranges? In many mammals and other endotherms, notable changes follow Bergmann's Rule, where body mass within a species gets larger at higher latitudes and cooler climates. It gets more complicated with ectotherms. Generally speaking, there is often a correlation between cooler temperatures and smaller maxiumum body sizes. One study noted: "poikilothermic giants on land become two - three times shorter per each 10 degrees of decrease in ambient temperature" (Makarieva, et al, 2005). However, another study (Olalla-Tárraga 2006) pointed out that no single environmental factor appears to explain the often inconsistent patterns of body size among snakes and lizards in northern temperate regions. Ashton and Feldman (2003) argued that snakes and lizards reverse Bergmann's Rule, while Ashton (2001) found that in two separate populations of the western rattlesnake (*Crotalus viridis*) group, one grew smaller in cooler areas while the other grew larger in cooler areas. What is interesting to me, is that among the maximum lengths given by Boundy (1995), several of the largest snakes come from northern portions of their ranges. The largest racer was caught in Connecticut, the largest eastern ratsnake was found in New York, the largest copperhead (1346 mm) came from New York, and a very large timber rattlesnake (1880 mm) came from Massachusetts.

When we begin to look at the giant snake stories, I think it will become obvious (at least to the ophidio-philes) that quite a few of the stories involve known species. Among the "typical" snake stories, likely culprits include ratsnakes, coachwhips, and, especially, racers. For some

Credit: Photographer unknown, vintage photo

stories, exaggeration of size plays its part. I have a strong suspicion, though, that as development in the late 1800s and early 1900s moved further into the rural areas, there was a genetic shift in some snake populations that accompanied the inevitable slaughter of our larger native species. Farmers would have been more careful toward the common ratsnakes, which they usually recognized as useful for reducing rodent pests, but other snakes were too often fair game. Rattlesnake hunts were often organized, but other snakes were targeted also, and as the largest snakes were killed, I suspect that this directly influenced the population genetics among our known native species. Also, the ecological landscape has changed drastically over the last century, particularly in the eastern United States. Habitat changes directly affect prey abundance, which acts as a limiting factor for maximum lengths.

So, while there may have rarely been twelve-foot long racers (just one example) in parts of Ohio one hundred years ago, the species may not be capable of reaching such lengths today. Then again, there may be regions where known species still have a genetic capability for extraordinary growth, but perhaps unrealized if environmental or other limiting factors work against that.

Now, from a strictly herpetological standpoint, a twelve-foot long colubrid would be very big, but not unreasonable. The largest colubrids known are in the Asian genus *Ptyas*, with *P. carinata* (formerly *Zaocys carinatus*) reaching 4000 mm (13 ft 1.5 in), and *P. mucosus* reaching 3700 mm (12 ft 1.7 in) (Manthey and Grossmann 1997). Several species of South American neotropical colubrids (cribos, *Spilotes*, etc.) also get fairly hefty. New maximum lengths for North American snakes continue to be recorded, so I would be very careful before making definitive statements regarding the potential maximum length that one of our native species could reach.

Credit: Will Stuart

The eastern coachwhip is a magnificent species: large, alert, and harmless (though will bite if handled). The photographer noted that this specimen exhibits typical coloration: black head and neck, turning to brown along the body and light tan towards the tail. Some coachwhips are completely melanistic. Ratsnakes and racers are often black also, which lead to the term, *black-snake*. Technically, *blacksnake* is not applied to a single species, so should be considered a generalized popularization (like *wildcat* or *black panther*), rather than a specific designation.

Pantherophis (Elaphe)

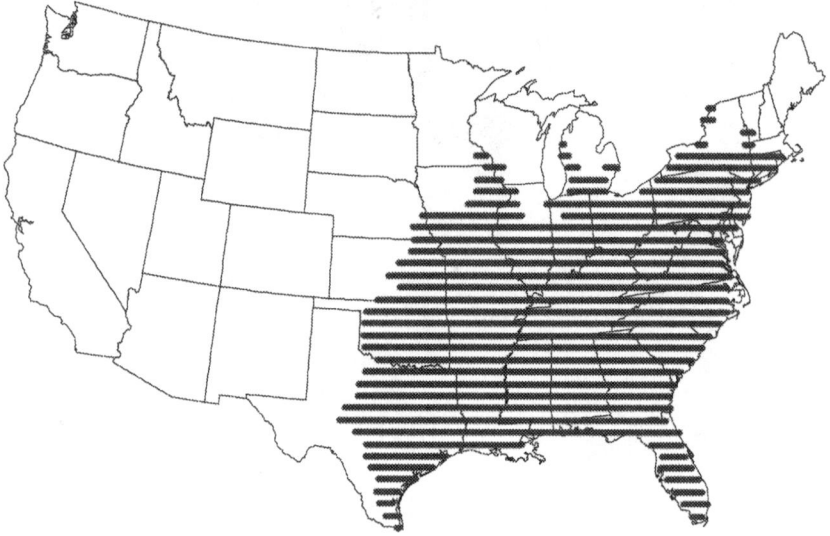

Until the last few years, many of the North American ratsnakes were considered subspecies of *Elaphe obsoleta*. Recent revisions based on phylogenetic evidence have made it difficult to keep up with the best scientific designations. Currently separated from the Old World *Elaphe*, our ratsnakes are now (mostly) in the genus *Pantherophis*, but even that may change in a few years, given discussion that ratsnakes should be placed within *Pituophis* with the bullsnakes and pinesnakes. The map above shows the wide range of the genus *Pantherophis* as it now stands. The eastern ratsnake, *Pantherophis alleghaniensis*, is the one species most likely to produce large specimens. (The former subspecies of yellow ratsnake, black ratsnake, etc., probably won't be redefined within *P. alleghaniensis*, but rather noted as clinal variations.)

Ratsnakes feed on rodents and other small mammals, birds, and bird eggs. Their activity period varies from diurnal to nocturnal depending on the season. They are great tree-climbers. Habitat varies to include deep woods, rocky forest hillsides, forest edges, swamps and bayous, grasslands, and farmlands.

Eastern (Black) Ratsnake
Credit: NBII / John J. Mosesso

Eastern (Yellow) Ratsnake
Credit: Zebulon Hoover

Coluber

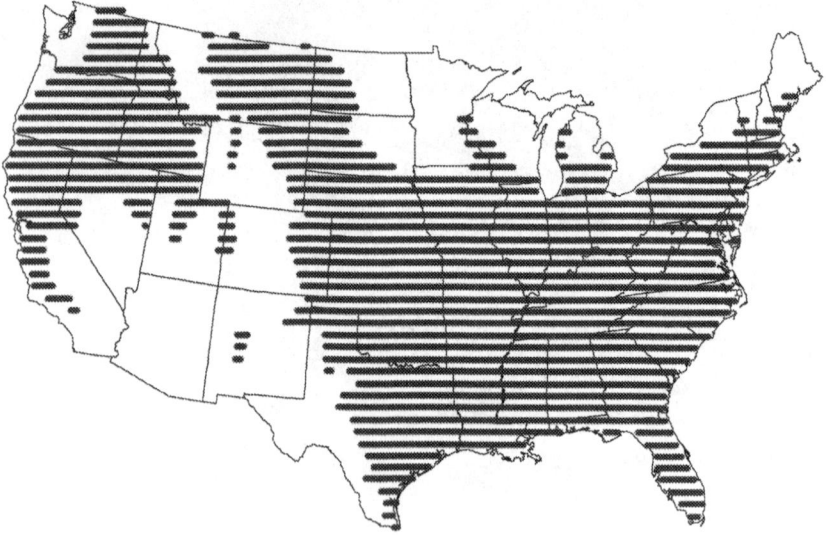

The racer, *Coluber constrictor*, has at least nine subspecies (more, depending on who you read). One former subspecies has been elevated to full species status, *Coluber mormon*, the western racer. This is another wide-ranging genus, with the largest snakes usually coming from the east. The northern and southern black racers show a coloration ranging from solid black to bluish gray, with a lighter underbelly. These are active, alert, diurnal snakes that feed on many different prey species (insects, amphibians, smaller reptiles, small mammals, birds). They take advantage of a wide range of brush, grass, and open forest habitats.

While this is a fairly common snake, never overestimate the average person's ability to not recognize native species. As the story that comes next shows, even a racer can make headline news. You'll also see that the snake, while large, is not beyond its maximum length. Newspapers didn't just focus on extraordinarily large snakes. There are hundreds of stories of "monster" snakes that only measured four or five feet in length.

Black Racer   Credit: Zebulon Hoover

Black Racer   Credit: Terry J. Alcorn

Woman Kills Huge Snake

Mrs. Anthony Bianco, 6539 Ashland ave., is shown displaying the six-foot snake which she killed Monday evening in her back yard by crushing it with the spade that she is holding. The reptile is believed to be of the blacksnake or blue racer type. No one is sure where the serpent came from, but it is thought that it may have escaped from a carnival man while he was stopping at a Southtown garage to have his car repaired.

—Economist Photo

Woman Kills 6-Ft. Snake at 65th, Ashland

Probably the largest snake that has visited Southtown for many years has been caught. And a spade, maneuvered by the white hands of a woman, has killed the black reptile.

Oozing forth through the tangled grass of her back yard, a large dark serpent, probably a blacksnake or maybe a blue racer, silently zig-zagged toward the home of Mrs. Anthony Blanco, 6530 Ashland ave., early Monday evening.

Neighbors who saw the snake in the yard were instantly frightened. Thinking that perhaps it was a copperhead, a deadly venomous serpent, none would step within striking distance of the reptile.

Calmly Kills Snake

Attracted by the commotion raised by the neighbors. Mrs. Blanco went to the back yard. Seeing the snake she cautiously walked to the side of the house and, taking up a spade which her husband had left leaning against the building, wielded it over the reptile and crushed its head.

Assured that the serpent was dead, the neighbors approached the reptile to measure its extraordinary length The snake was more than six feet long. Upon closer examination it was thought that the serpent belonged to the copperhead family, but snake experts at the Field Museum, advised of its description, explained that it probably was a blacksnake or blue racer.

Copperheads, they stated, do not attain a length of more than three feet, and inhabit only the eastern states.

Reptile Story Spreads

When the story of the huge snake that Mrs. Blanco had killed spread about the neighborhood, it was recalled that last Sunday morning a carnival man had stopped at the F. and D. garage. 72nd pl. and Ashland ave., to have his car repaired.

No one knows who the carnival man is, but in the rear of his auto mobile, tucked among the folds of a canvas, a large, dark snake was discovered by the repair man.

It is not known for certain whether this snake escaped from the automobile and eventually wound its way into Mrs. Blanco's yard, for the carnival man and his auto have disappeared.

But neighbors still contend that it is most likely that the carnival man's black serpent was killed early Monday evening by the white hands of Mrs. Blanco.

Chicago, Illinois, *Southtown Economist*
July 11, 1928

Masticophis flagellum

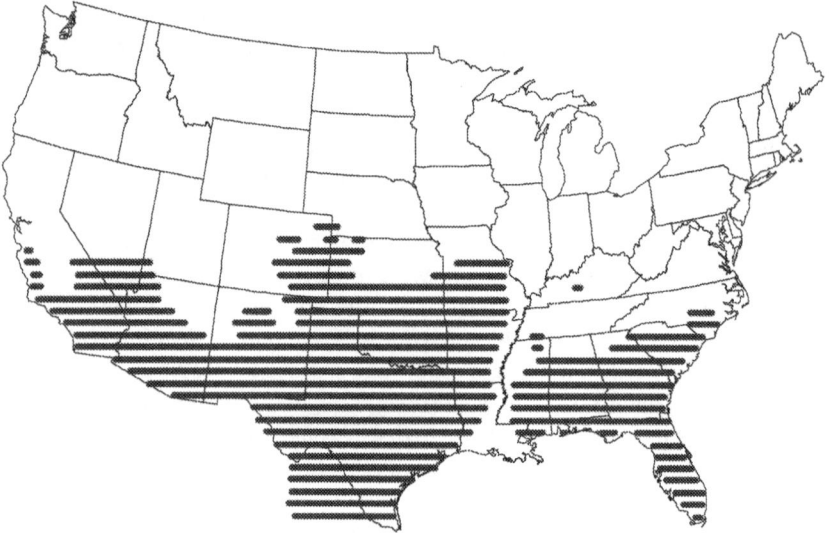

While there are other species of *Masticophis*, only the large *M. flagellum* is a good candidate for giant snake stories. There are seven subspecies. Coloration varies even with a subspecies. The eastern coachwhip usually shows a black head fading away to a light tan towards its tail, but it can also be melanistic or even completely pale. In several western subspecies, red or pink individuals are not uncommon, and blotching is occasionally seen.

Coachwhips are diurnal, but may only be active in mornings or late afternoons. They inhabit grasslands, scrub, open forests, rocky hillsides, and desert regions. Like the racer, it is a generalist, feeding on a wide range of warm-blooded and cold-blooded prey.

The braid-like pattern on the tail lends to its name, and early folk-lore spoke of coachwhips catching unwary victims and flogging them unmercifully. That, of course, doesn't happen, though if you pick one up, it is liable to bite you repeatedly.

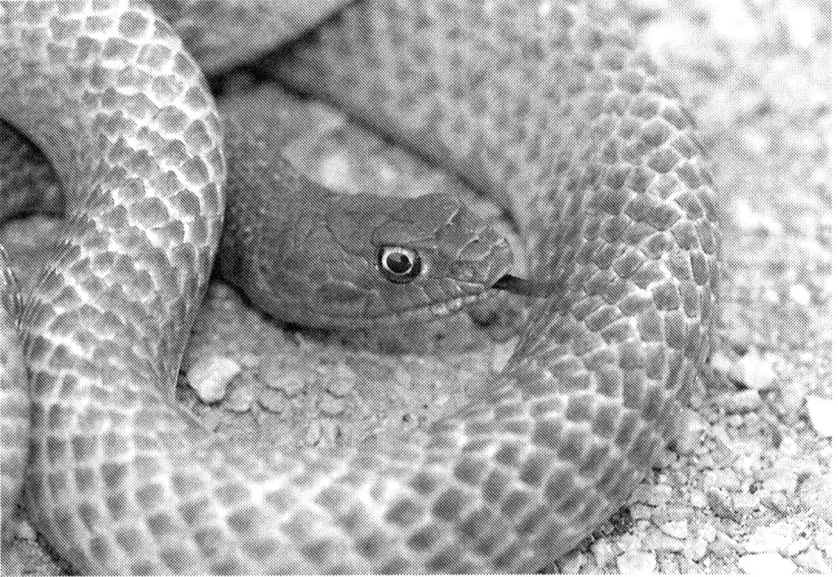

Western Coachwhip   Credit: Rusty Dodson

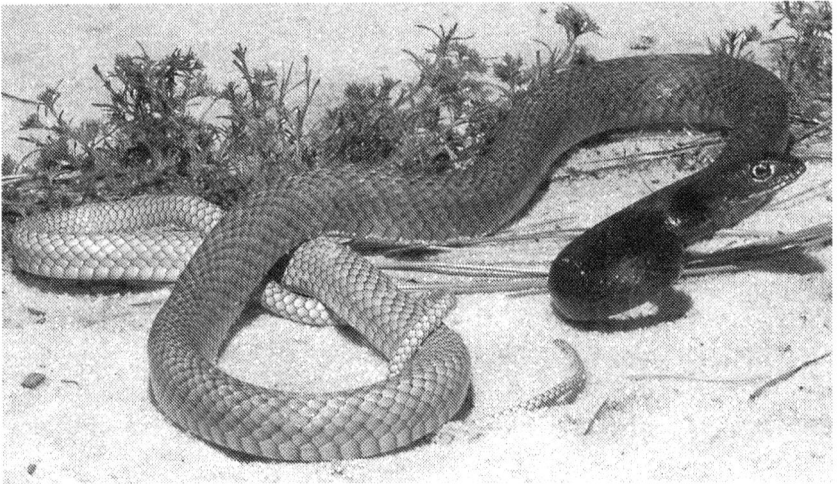

Eastern Coachwhip   Credit: Zack Bittner

Pituophis

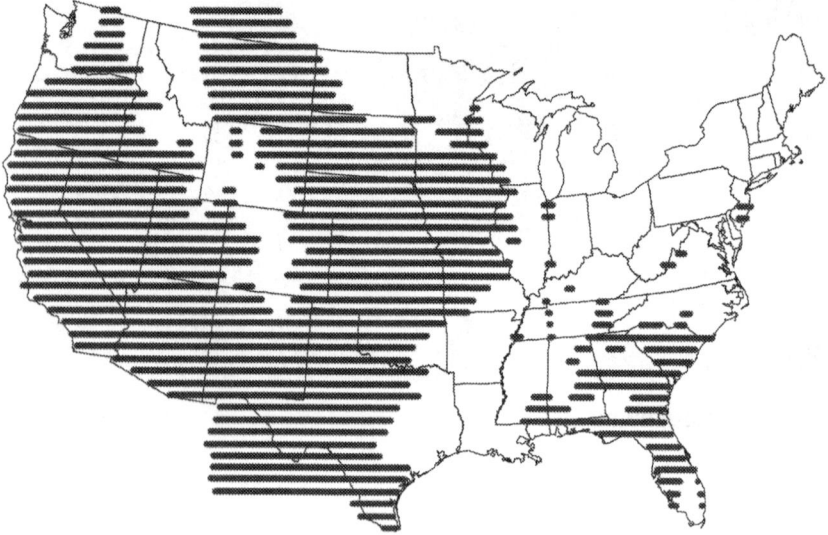

Snakes in the genus *Pituophis* are bulkier than the slim coachwhips and racers, terrestrial, and often fossorial. Several species are cryptic in their habits, and can be difficult to locate during normal fieldwork. *Pituophis catenifer* is a habitat generalist, while *P. melanoleucus* is usually found associated with the dry sandy soils of pine-oak forests. Small mammals make up the bulk of their diet, with the occasional bird, bird's egg, or lizard thrown in.

Besides the verified 8 ft 9 in specimen (Devitt, et al, 2007), there is a reported bullsnake of 9 feet in length, but that was a skin-only specimen rather than a live animal, so the possibility of stretching has to be taken into consideration.

As I have noted previously (Arment 2004), I consider *Pituophis* to be a good candidate for the Broad Top big snake reports from central Pennsylvania. Not only does the coloration, behavior, and habitat fit the *Pituophis* profile, but the only known specimen of *Pituophis* in Pennsylvania—late Pleistocene material from a Bedford County cave—comes from that very region (Holman 2000).

Bullsnake (*Pituophis catenifer sayi*)   Credit: Rusty Dodson

Northern Pinesnake (*Pituophis m. melanoleucus*)   Credit: Kasi Lodrigue

Drymarchon

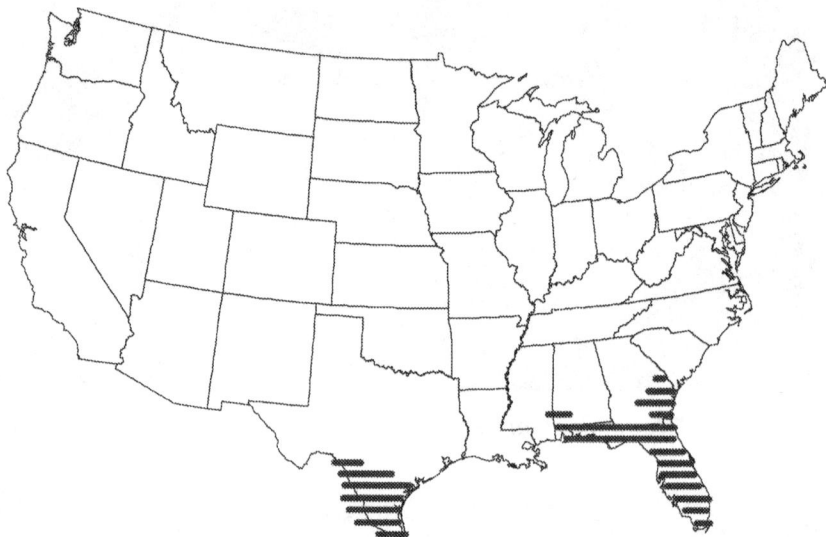

Indigo snakes are large terrestrial (often fossorial) snakes. The eastern indigo, *Drymarchon couperi*, may vary its habitat use seasonally, from sandy areas to wetlands, while the Texas indigo, *Drymarchon corais erebennus*, is found in mesquite brush, sandhills, and grasslands close to water sources. They are diurnal and feed on a wide range of prey species (warm and cold-blooded).

There is a report of an eastern indigo reaching 9 ft 2 in, but I have not found a reliable source for that measurement. Another subspecies of *Drymarchon corais*, the yellowtail cribo of Central America, is known to reach 10 feet in length.

Historically, the eastern indigo was a common snake for snake charmers and traveling menageries, due to its large size and docile temperament (when tamed). There are news accounts of carnival escapees of "big black snakes" which almost certainly refer to this species. Today, our native indigos have legal protection, though there are numerous captive-born indigos in the pet trade.

Eastern Indigo Snake   Credit: USFWS / Pete Pattavina

Texas Indigo Snake   Credit: Karri Egger

Crotalus

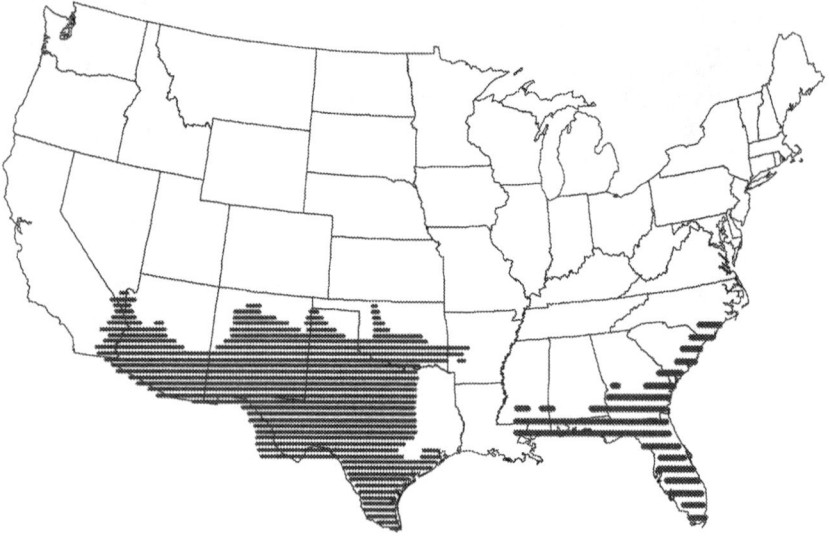

West: *Crotalus atrox* / East: *Crotalus adamanteus*

The largest rattlesnakes in North America are the eastern diamond-back, *C. adamanteus*, and the western diamondback, *C. atrox*. Rattle-snakes have traditionally been the "vermin" most often culled over the last few hundred years, and hunting has probably both reduced the genetic potential within the populations and lowered the overall life expectancies, which contributes to a decrease in size. Christman (1975) noted that evidence suggests "large specimens of *C. adamanteus* were probably more common in the past than now." Environment also has its say, and there have been some very large eastern diamondbacks raised in captivity.

Years ago, some Pleistocene material in Florida was named *Crotalus giganteus*, but further examination determined that it was no larger than eastern diamondbacks are known to grow, so *C. giganteus* was synony-mized (Christman 1975).

Eastern Diamondback Rattlesnake   Credit: C. Emory Moody

Western Diamondback Rattlesnake   Credit: Joel Johndro

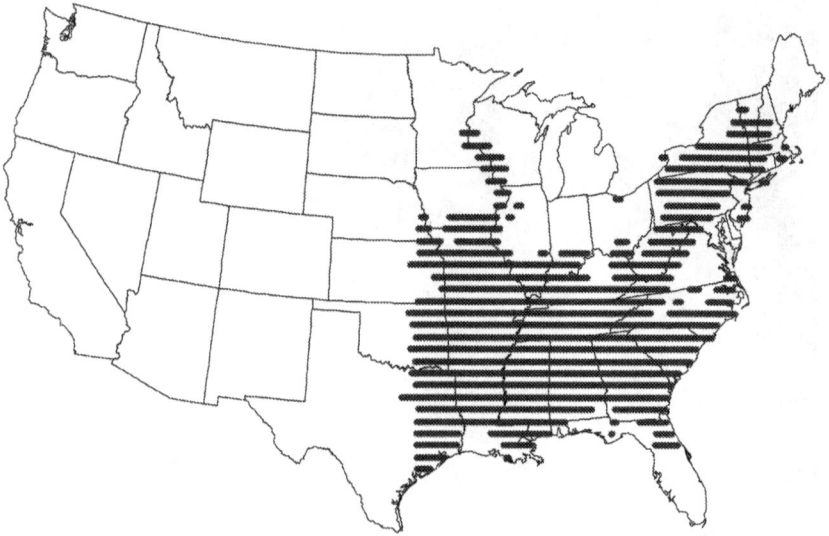

Range of *Crotalus horridus*

There are 16 species of rattlesnakes (*Crotalus* and *Sistrurus*) in North America, with a greater number of subspecies. Most are small, and I would suggest caution when evaluating reports of giant rattlesnakes outside the known ranges of *C. atrox* and *C. adamanteus*. Several reports come from the northern range of *C. horridus*, and there is a very intriguing account from West Virginia that I think is worth further investigation. In the western states, though, the prairie rattlesnake (*C. viridis*) - western rattle-snake (*C. oreganus*) complex really doesn't offer a likely source for big rattlesnakes.

Timber/Canebrake Rattlesnake   Credit: Kevin Batchelor

Prairie Rattlesnake   Credit: Rusty Dodson

Boids

Traditionally, the most common exotic boas and pythons to be found roaming loose have been the Burmese python (*Python molurus bivittatus*) and *Boa constrictor*. Being large, attractive snakes that are usually docile when raised from a juvenile, these have been kept in North American zoos, circuses, and traveling menageries since the colonial period.

It's no surprise then, that these two snakes have now entrenched themselves in southern Florida, as discarded pets founded populations in a brand new habitat. The Burmese python has particularly been active, and certainly receives a great deal of press. It was introduced into the Everglades in the 1990s, and the population appears to be rapidly increasing. You might call it the "perfect storm" of feral introductions. *Boa constrictor* was introduced far earlier, as will be seen, but apparently never took off. Recent sightings, including one of a pregnant female, suggest that it is now established, but more research is needed on how stable the population is.

Today, though, there are many more species that could potentially be the identity behind a "big snake" sighting, as herpetocultural advances offer reptile hobbyists a wide range of large boids as pets. An unsecured cage provides ample opportunity for a snake escape, and we are beginning to see more exotic species show up in unexpected places. (And, of course, there are the occasional idiots who deliberately dump their big snake into the woods rather than locating a reptile rescue, when they get tired of taking care of a large snake.)

Here are a few of our modern-day big snake stories:

August 1989: A 20-foot long, 250-lb reticulated python was dug out from under a Fort Lauderdale house, where it appeared to have been residing for years. (*Pacific Stars and Stripes*)

February 2004: A 15-foot Burmese python was shot and killed in a Crawford County, Kansas, strip pit by campers. (Pittsburg, KS, *Morning Sun*)

August 2006: A 19-foot Burmese python found dead in the White River, Indiana, caused quite a stir in the community. It turned out the snake had died of natural causes in captivity, and was

Burmese Python   Credit: Photography by Varina

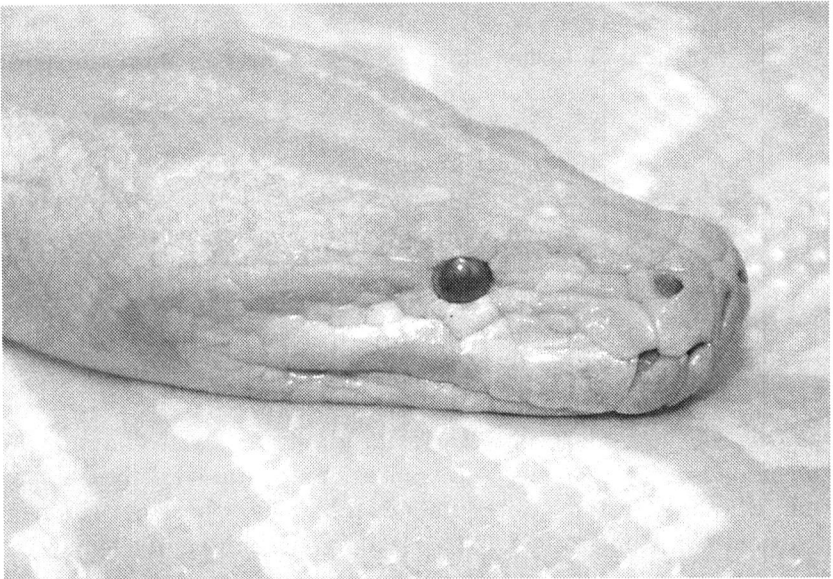

Albino Burmese Python   Credit: Vladimir Mucibabic

kept in a freezer for future taxidermy. Relatives of the snake's owner "borrowed" the freezer without informing him, found the snake in it, and dumped it in the river. (Indianapolis *Star*; Noblesville, IN, *Ledger*)

June 2007: A Philadelphia suburb was in a frenzy over reports of a 10- to 12-foot snake that bit the heads off kittens. (For those unaware, snakes don't actually do that.) (CBS3.com)

June 2007: The shed skin of a 10- to 12-foot Boa constrictor was found in the woods of Patrick County, Virginia, by a V-DOT worker. (WSLS.com)

September 2007: A frantic Burr Ridge, Illinois, resident videotaped what she and her family believed to be a 10- to 15-foot *Boa constrictor* on her driveway. This, of course, lead to neighborhood panic. A USFWS officer viewed the tape and pointed out that it was clearly a native species, the fox snake, and probably not more than 4 or 5 feet in length. (Chicago, IL, *Tribune*)

November 2007: Several homeless people in Stockton, California, reported seeing a 15-foot snake near a waterway, Mormon Slough. Investigation turned up the shed skin from an approximately 6-foot *Boa constrictor*. (Stockton, CA, *Record*)

December 2007: Two duck hunters in the Metzger Marsh State Wildlife Area, Lucas County, Ohio, came across a seven-foot python in the weeds. It was sluggish, but still alive. They gave it to a local reptile enthusiast. The newspaper identified the snake as an "African rock python," but the accompanying picture of the snake showed that it is not that species, but a carpet python. Carpet pythons are from Australasia, and common in the pet trade. (Toledo, OH, *Blade*)

Obviously, modern day stories of giant snakes are bound to be convoluted with exotic releases/escapees. It's a natural folkloric progression, where "monsters" of the unknown have now become "monsters" threatening the environment. This leads to something of a conundrum when evaluating reports, as we now have multiple possible explanations (of varying probability) for boid-like giant snake stories.

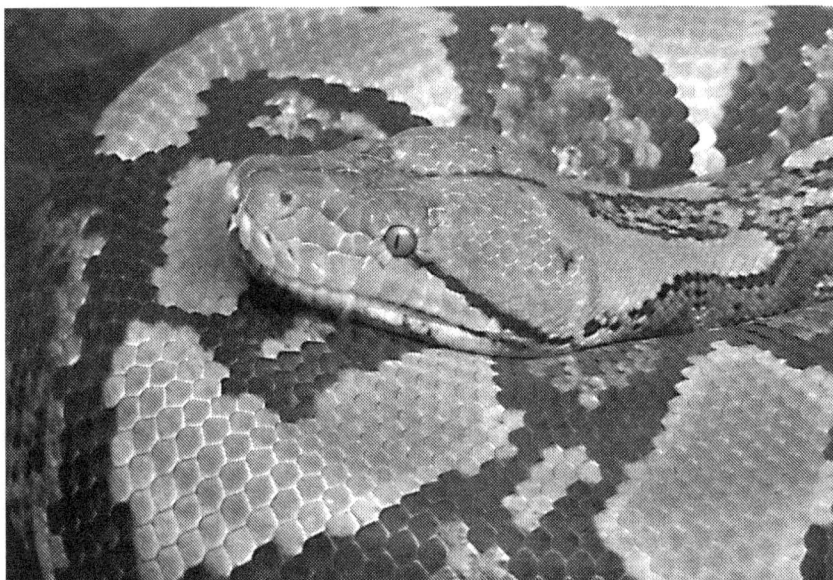

Reticulated Python   Credit: Anastasiya Igolkina

African Rock Python   Credit: Lucian Coman

1. Exotic species being mistaken for undescribed species.
2. Native species being mistaken for undescribed species.
3. Undescribed species being mistaken for exotic species.
4. Native species being mistaken for exotic species.

You'll see that points 1, 2, and 4 do in fact occur. So, is it possible for an undescribed species to be mistaken as an exotic? I've noticed a lot of shady identifications in news articles, and not all "experts" chosen by news reporters truly know what they are talking about. The python invasion in the Everglades particularly presents a difficulty; given that there were reports of "big snakes" in the Everglades as far back as Willoughby's account in 1910, what chance would a rare native (but undescribed) species have against the invasive Burmese python? I suspect not much. Oddly enough, Ferriter, et al, (2006) even noted the Willoughby stories, but brushed them off as "rare and infrequent circus animal escapees," which is an absurd pronouncement on those Seminole legends. The trick for investigators will be walking that line between recognition of exotic releases and the possibility (however small) of an undescribed species.

With at least 45 species of reptiles and amphibians introduced into Florida (Meshaka 2006), there has been a great deal of discussion over how to stop other large species from spreading into the Everglades and other regions. Reed (2005) analyzed the potential risk of 23 species of pythons and boas, and determined that three species are most likely to become invasive problems: *Boa constrictor*, the reticulated python, and the ball python. Of these, the reticulated python is the most worrisome.

My suggestion for investigators is to a) become familiar with the species of boas and pythons that are common in the pet trade, and b) recognize the folkloric basis behind communication of stories, even in newspapers, so that you can best determine what the actual facts are and what is opinion or speculation masquerading as fact. Species that cryptozoological investigators should be familiar with include the *Boa constrictor*, *Eunectes murinus* (green anaconda), *Eunectes notaeus* (yellow anaconda), *Python molurus* (Indian python), *Python reticulatus* (reticulated python), *Python sebae* (African rock python), *Morelia amethistina* (amethystine python), and *Morelia spilota* (carpet python).

Other species are either too small (ball pythons, spotted pythons), too infrequent (olive pythons, papuan pythons), or too expensive (Boelen's pythons, black-headed pythons) to be likely culprits behind the average big snake story today.

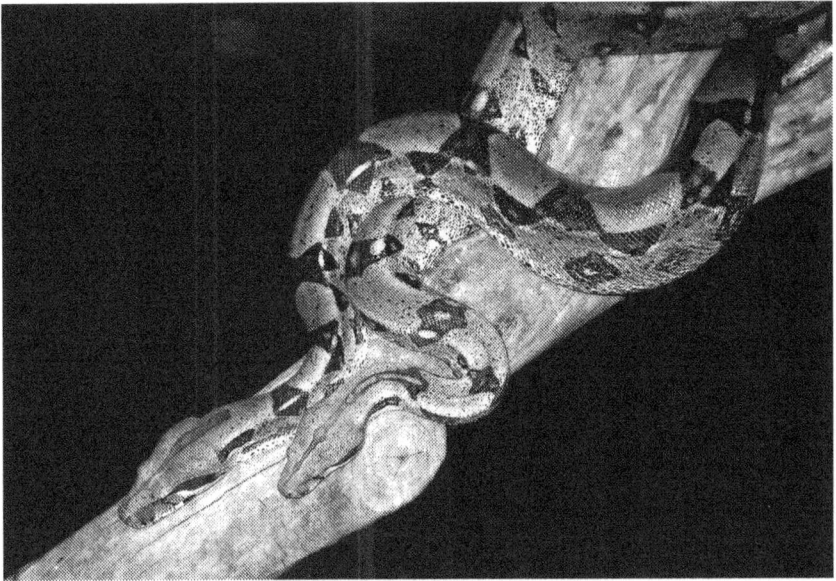

Boa constrictor   Credit: Sascha Burkard

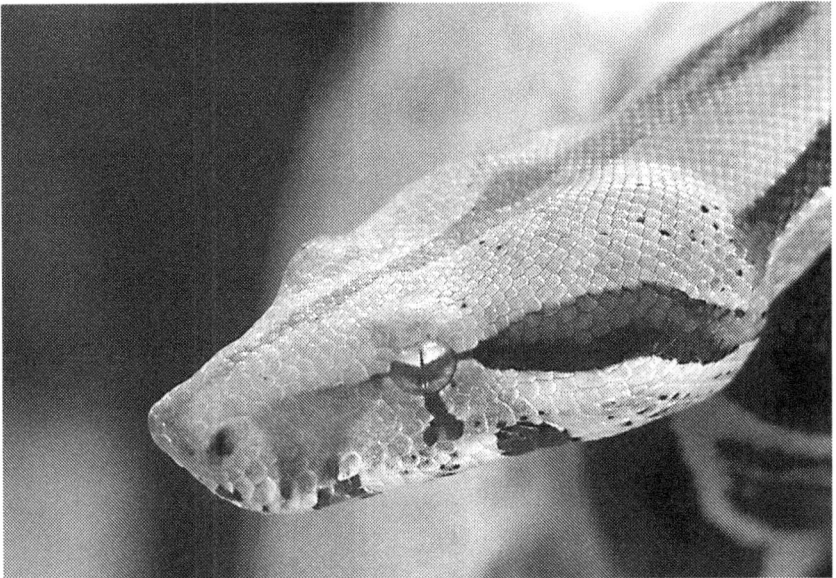

Boa constrictor   Credit: Peter Ong

Here is one case of a big snake from central Pennsylvania that is clearly an out-of-place exotic. Despite the newspaper's suggestion, this is a *Boa constrictor*, as the tail markings are distinctive.

Shades Of John Barleycorn! !

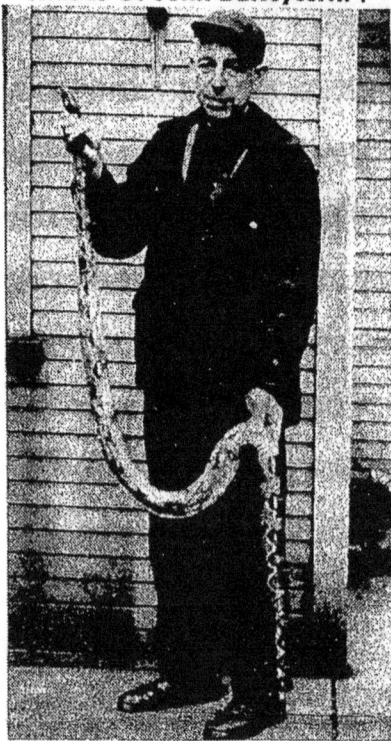

Game Protector Ross Metz holds a seven-foot long reptile, believed to be an Indian Python, which was found and turned over to him by James Patterson of Alexandria.

Mr. Patterson, while walking last Tuesday along a stream in the area known as "The Meadows," between Duncansville and Altoona, saw the snake's lifeless tail sticking out of a driftwood pile along the bank, and hauled it out, foot by foot.

The snake definitely is not common to Pennsylvania, and Game Commission officials think it must have escaped from a traveling circus or carnival and perished from the cold.

Huntingdon, PA, *Daily News*,
February 14, 1953

The Potential for
Unknown Species

Having examined the accounts that follow, I am not optimistic about the existence of a large terrestrial snake that is completely unknown (not to be found, let's say, within one of our native genera). One hypothesis we can dismiss immediately is the idea that a single unknown species is responsible for reports from all over the continent. There is no underlying pattern of physical, geographical, or ecological characteristics that suggests that a single species of giant snake is responsible for the bulk of accounts. Coloration and pattern, behavior, and habitat are too diverse among the reports.

We can also note two other problems with this idea. First, where there are big snakes, there must be little snakes—many little snakes. If someone argues that an unknown species of giant snake inhabits a range that includes Indiana, Kansas, New York, Maryland, and Georgia, we would reasonably expect that someone, somewhere, would have captured and reported a strange juvenile snake. Snakes have a wide and enthusiastic amateur and professional following that includes field herping and hands-on captures in all kinds of habitats, throughout most of North America. While I have no difficulty with the idea that certain habitat "pockets" may be overlooked by herp enthusiasts, and some areas are certainly not as well surveyed as others, there is no good reason to expect a wide-ranging snake to be overlooked. Anyone with a rudimentary knowledge of snakes can distinguish our native species, even in juveniles that sometimes share common coloration or patterns. They are just not that difficult to identify.

Second, there is the problem of overscaling the importance of data points in creating such a wide distribution. Ten to fifteen years ago, there were perhaps a few dozen reports of North American giant snakes in the cryptozoological literature. Of these, a fair number can now be identified as inaccurate or false, while almost half were just generic lake monster

73

sightings that incorporated snake-like metaphors. You'll notice that in this text I only include aquatic reports when the sighting appears to actually involve a snake. I do not include lake monster accounts of animals that a) lift their heads significantly out of the water, especially while swimming, b) have "horse-like," "camel-like," or "sheep-like" heads, or c) include other physical characteristics or behaviors that are not found in snakes. I won't dismiss the idea that there may be an unknown species behind some lake monster sightings, but most are not consistent with the description of a snake. So, early attempts to "fill in the blanks" with inaccurate data were most responsible for the poorly-supported concept of a wide-ranging unknown giant snake.

For those stories that truly suggest that very large snakes was seen (where falsehoods, misidentifications, and exaggerations are not immediately apparent), it looks to me as if they are best explained by:

1. Historically larger maximum lengths within populations of recognized species.
2. Out-of-place exotics (which would particularly explain "single-season wonders," dying off when colder winter temperatures arrive).
3. Small localized populations of phylogenetically distinct snakes (i.e., larger than average), closely related to our recognized native species. They may not show species-level morphological differentiation.

(Let me take a quick rabbit trail, and suggest 2b: the historical introduction of an exotic species, specifically *Boa constrictor*, through Native American trading, into California. Given that several species of live birds and reptiles were traded in the pre-colonial southwest, I can potentially see how boas might have ended up in early California—and, may have been extirpated by now, if they ever existed. We know that some exotics are successfully established in that state, including at least one colony of chameleons, accidentally released by officials checking on a reptile breeder. As several of the California stories involve boa-like descriptions, might that be a possible explanation? There's no evidence, but I'd keep it in mind.)

But what about this possibility?

4. Small localized population(s) of one or more unknown species, not closely related to known species.

This is a possibility, though it probably isn't as likely as some cryptozoological researchers might suggest. In and of itself, size is not the best indicator for detection of a snake species (Reed and Boback 2002). Snakes that are nocturnal, burrowers, sedentary (as in ambush predators), or well camouflaged will be less likely to be noticed, regardless of their size. Geography also is important, as regions with poor access or little traffic (particularly by individuals capable of discerning that a species is remarkable), may certainly hide something new.

The possibility of a large unknown species raises a few questions:

Is there any fossil evidence pointing to undiscovered snakes?

Basically, no. Documentation of Pleistocene snake material (particularly from caves or sinkholes) is actually fairly thorough. Even the smaller, cryptic species seem well covered. I've not seen any records for large genera that disappeared, or of significantly larger fossil specimens of modern species.

Are there earlier fossil snakes in North America that were larger? Sure. Large fossils of the aquatic *Palaeophis* and *Pterosphenus* are known from eastern North America (some species of which could reach thirty feet in length), and there are some early terrestrial boids from various points on the continent. (From what I've seen, most of the latter appear to have been in the 4 to 8 foot range.) Without recent fossil material, however, it's just fanciful speculation to suggest that any of these are still around.

Can large snakes live in temperate climates with high seasonality?

This shouldn't be a major limiting factor. Go too far north, and yes, there is a far shorter season available for growth, but for most regions in North America, I don't see any evidence that a temperate environment is a barrier for size. Large snakes that live in temperate climates may modify their behavior for optimal thermoregulation and foraging, as do some populations of carpet pythons in Australia (Ayers and Shine 1997).

How big can snakes get?

Darren Naish (2007) briefly reviewed giant snakes. It looks like 10 meters (+) is still about the maximum confirmed length for living boids, while some fossil species may have reached 15 meters or more in length. (Extrapolation is required on fossils when only a few vertebral fragments are known, so there is no way to be absolute on the largest suggested lengths.)

I have not seen any evidence of fossil colubrids reaching lengths greater than living species, but would be very interested in hearing about research on such.

Giant Snakes by State or Province

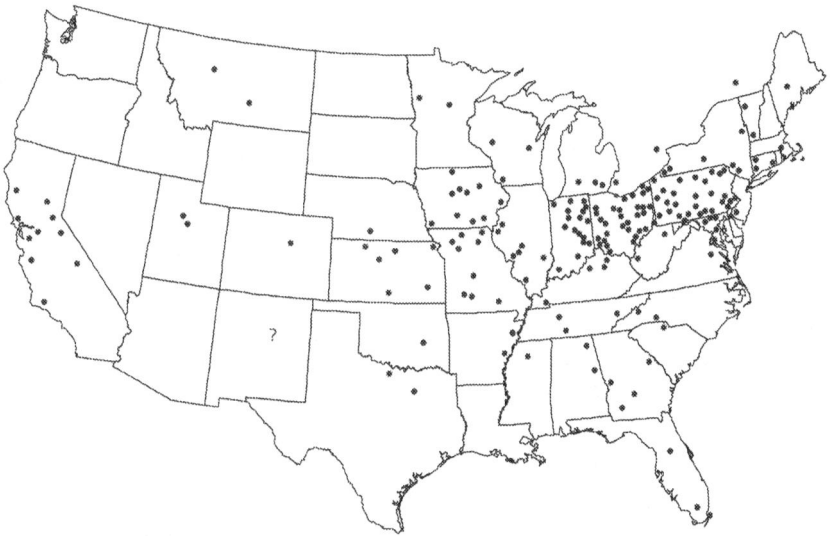

This map is offered with caveats. It shouldn't be taken beyond face value, although I do think it makes an interesting point. Too often, though, rather than being content to offer informative data (showing where sightings have occurred, showing numbers of recognized sightings, etc.), some individuals in cryptozoology create maps and use them to mask insupportable conclusions or draw inappropriate correlations. Maps are tricky things. Even in "normal" zoology, there is debate over how to draw and interpret static biogeographical range maps of known species, whose true distribution is in constant flux.

This map is specific, noting only the locations mentioned in the following accounts of alleged giant snakes. It does not display the relationship of locality to number of reports—a location may have a single report or dozens. The map also does not distinguish between possible and improbable accounts—as discussed earlier, some reports will most likely fit known species, while others may point to an unknown species or a fictitious one.

The map does, however, suggest that as far as locations are concerned, the distribution of alleged giant snake reports is not random. Some locations with many, many snakes do not appear to have as many (if any) reports of really big snakes. If a major factor in giant snake sightings is a combination of ophidiophobia and exaggeration of length, I would expect to see the distribution to generally follow the ranges of some of our larger snakes. And, would a distribution created mostly from newspaper hoaxes center on Indiana-Ohio-Pennsylvania as it currently appears? This does not mean that the presence of an unknown species is the only possible factor influencing the distribution of localities. Cultural influences can be subtle, but effective. Also, I cannot be positive that my collection of historical news accounts reflects the true distribution of sightings of big snakes across the continent. Because of the manner in which reports were shared by different newspapers, there is at least some coverage from most parts of the country. It is probable, however, that some rural areas with giant snake reports have had little news coverage or distribution.

The most important use of the data is to determine which areas are most suitable for future investigation. A good location can be key to uncovering new sighting reports, which allows us to better prepare for fieldwork and physical evidence gathering. If nothing else, I hope that this book spurs further field investigation into a fascinating, but often overlooked subject.

Alabama

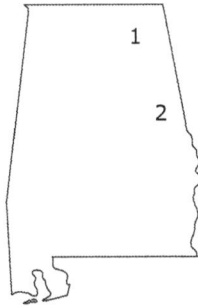

1. Kilpatrick, Marshall County
2. Randolph/Clay County Line

The primary account of a giant Alabama snake on the following pages, from 1957, suggests that big snake folklore was historically present in at least the northern region of the state, but this will require confirmation through other sources. The only other account I've seen is the story of two men, driving along the county line between Randolph and Clay counties in 1977, who saw a "tree"-sized snake, dark with light markings in a fish pond alongside the road (Smith 2004). They watched it swim across the pond and disappear into its depths.

Big Snake in Alabama
Reptile Said to Be 30 Feet Long

Guntersville, Ala., July 15 (AP)—Reports that a 20 to 30-foot long snake has been seen on a Kilpatrick community farm are bringing hundreds of curious visitors to this north Alabama section.

One service station has put up a sign saying "snake information here."

Dan Langston says he saw the snake on his farm a few days ago. He described it as "at least 20 feet long" with a wide, flat head.

Several residents said there has been talk of a large snake having been seen in this section for years.

No snakes that large are native to Alabama. There have been no reports of reptiles of great size escaping from zoos or traveling menageries in these parts.

The farmer insists that more people have seen the snake then have said anything about it. He asserted that many people are scared to talk for fear no one will believe them.

Recently it was reported that the snake had moved from this Marshall county area into the edge of DeKalb county. Several score farmers from DeKalb county decided to try lo get the snake back into Marshall county, where they feel it belongs. They went hunting for the "monster" armed with guns and long sticks, but didn't find anything.

This group tried to go upon Langston's farm but he said he didn't want anybody hurt and stopped them.

Law enforcement officers generally are skeptical. Langston blames the disappearance or death of four cows on the monster. He said the dead cow had marks as if it had been squeezed to death, as the large snakes kill their victims.

Lowell, Massachusetts, *Sun*
July 15, 1957

Alabama's 'Monster'

The countryfolks of Marshall County, in north Alabama, are a right smart exercised by rumors of a monster in the form of a snake that several people claim to have seen. Farmer Dan

Langston of Kilpatrick blames the disappearance of three cows and the death of a fourth on the reptile. He said the dead cow had evidently been encircled and squeezed to death.

Folks in adjoining Dekalb County got wind that the "monster" had abandoned Marshall County and invaded Dekalb, so they organized a posse armed with guns, ropes and "l-o-o-n-g" sticks to search for the critter. But Brother Langston stopped them at his property line; said he didn't want anybody to get hurt.

Law officers scoff at the whole thing. They say no snake of the constricter type has escaped from any circus or zoo anywhere near Marshall County. Those who claim to have seen the thing claim it is 20 to 30 feet long and as large around as a telephone pole.

Well, as one dispatch points out, it may just be the heat, but you'd have a hard time convincing those who have "seen" the snake that the weather had anything to do with it.

The Fort Worth-Dallas area twittered for several days not too long ago over a snake story, but that one was real enough. The snake had escaped from a zoo, and was eventually rounded up. Earlier Oklahoma City had its leopard scare, and sooner or later nearly every community is titillated by the strange and the bizarre.

We remember that these scare yarns were much more common fifty and sixty years ago than now. People didn't have much to entertain themselves with in those days, and anything of a hair-raising nature was welcome. Nowadays they have TV, radio and riots on the baseball diamond.

Abilene, Texas, *Reporter-News*
July 16, 1957

Midsummer Madness

Those who follow the pattern of news know there is a definite order in the way fantastic and fabulous stories emerge. They are more likely to prevail when it is hot than when it is cool. They occur more frequently when major news is scarce than when it is plentiful.

So it is no surprise to hear in these days of midsummer doldrums that an Alabama farmer says he has seen a snake between 20 and 30 feet long, big around as a telephone pole, and which may have killed several cows. What is more, by way of decor the critter is said to sport big brown spots.

Although a posse of Alabamans was out looking for this mammoth snake, this report fits nicely into the family tree of Paul Bunyan yarns, accounts of the abominable snowman of Tibet, the perennial sea serpent, ghost ships and the wolf boy somewhere amid the tundras or steppes of Asia.

Fresno, California, *Bee Republican*
August 20, 1957

Arizona

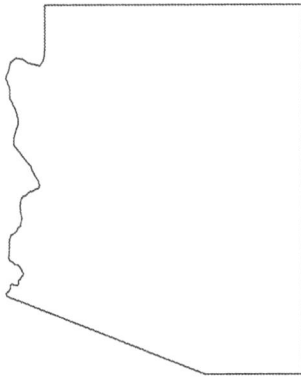

I don't have any specific reports from Arizona, but will note that there are legends of large water serpents among the Hopi, Navajo, and other Native American groups.

In 1957, a dead 18-foot python was found in Tucson's De Anza Park. It turned out to be a practical joke by the owner of South Dakota's Reptile Gardens, who owned an Arizona tourist cavern. (Hayward, California, *Daily Review*, February 4, 1957)

Arkansas

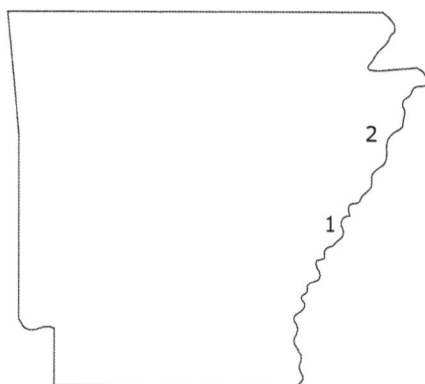

1. Helena, Phillips County
2. Turrell, Crittenden County

In June of 2005, there was a mention by KAIT Channel 8 that two "yellow anacondas" had been seen in the Wapanocca National Wildlife Refuge near Turrell. This is one of those cases where we have to seriously question the identification of a reported snake. A fisherman reported seeing one in 2004, stating that it was as long as his 14-foot boat and thick enough that two hands wouldn't be able to go around it. A second snake, about six feet long, was reported by a wildlife official in Spring, 2005. It is *highly* unlikely that yellow anacondas could overwinter successfully that far north, and given that no descriptive details were given, we have no way to determine what was actually seen.

Life in Arkansas

by Bob Harding
Associated Press Staff Writer

There's a "monster" snake in Storm Creek Lake near Helena. No one has seen it yet this year, apparently, but several folks saw last year.

Mrs. James Mayfield of Helena saw the monster while she and her mother, Mrs. Ann Fletcher, were fishing. If was about 12 feet long, wrapped around a log, she said.

"We fish at Storm Creek Lake often and are accustomed to seeing snakes and turtle," she added, "but we had never seen any thing to compare with this."

At first, she said, people "laughed at us and said we had just been fishing too much."

But Drs. A. A. Slutzky and Henry D. Grochau of. Marked Tree saw the thing, too. They said it was 9 to 12 feet long, big around as an auto tire and dark gray with black markings. They thought it might be a boa constrictor, a tropical snake which grows up to 12 feet in length.

Dr. James Wise of Marvell lends support to the boa theory.

He reports that in 1957 he accompanied Dr. Henry Farris of Jacksonville, Fla., to Storm Creek to turn a boa loose.

He said Farris had the reptile when both men were medical students.

Clarence Taylor of the Pine Bluff *Commercial* is one man who's quite interested in the "monster." He's urged Storm Creek fishermen to keep a sharp eye for the creature this summer, and to take a camera along because "a good clear, sharp picture of the 'monster' would be most welcome."

Blytheville, Arkansas, *Courier News*
May 5, 1960

California

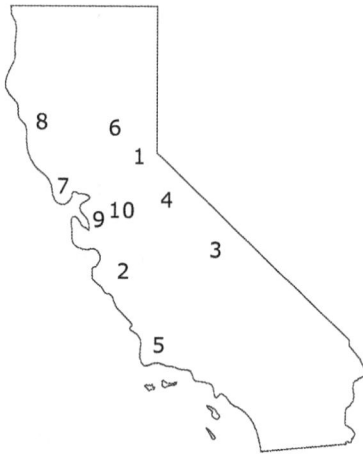

1.Placerville, El Dorado County

2. Tres Pinos, San Benito County

3. Diablo Hills, Inyo County

4. Calaveras County

5. Santa Barbara County

6. Yuba river

7. Sonoma County

8. Ukiah, Mendocino County

9. Decoto, Alameda County

10. Martinez, Contra Costa County

A monster serpent was seen recently three miles outside of Placerville, and described as red in color, crossed with black, and about forty feet in length. When it was first seen it was lying in a hollow on the side of a hill, its jaws distended and its tongue protruding; the last bifurcated like most of the serpent kind. It rose soon after being seen and moved off into an adjourning wood, where it was lost to sight. Its motion was slow and undulating, not sidelong, like the movements of lesser serpents; its immense weight, however, crushed everything beneath it and left a track in the grass similar to the swathe of a mower.

Reno, Nevada, *Nevada State Journal*
July 16, 1874.

A Gigantic Snake Story.—While one of our sportsmen was hunting lately in the Tres Pinos country he killed a fawn and left it on the ground to go in pursuit of other game. Returning a short time after, he found to his amazement and horror that a huge serpent was coiled around the body. Judging from the mischievous look of the reptile's eyes, the nervous curving of his neck and the threatening poise of its elongated head that it would not surrender the prize without a fight, the hunter concluded to retreat and take aim at the glittering thief at long range. After seating himself behind a rock a hundred feet distant and getting his nerves somewhat steadied for work, he blazed away, and spent about twenty charges of treble B shot without any visible effect. Feeling his pulse, he concluded from its rapid run he must be nervous. The snake by this time had uncoiled itself from the carcass and had twisted itself in the form of a corkscrew ready for a spring. With a manly effort our hero controlled himself and gave a firm and steady pull at the trigger. The serpent's head was nearly severed from its body by the shot and its form was at once relaxed and stretched out prone and powerless upon the ground. When the hunter considered it safe he advanced to get a closer view of his victim, and in stepping over the ground it measured twenty feet in length. The body was as large around as a child's waist. While looking out and wondering to what genus of large serpents it belonged he was horrified to see about 120 small snakes twisting, hissing and emerging from a nest close by. Each

particular hair of his head stood on end, and his legs were suddenly electrified with wonderful motion as he skedaddled like a
racehorse from the scene. There is a nonvenomous serpent of the
boa family in California, but it attains full growth only in warm
climates. This one was of uncommon size, and some of its off-
spring may yet attract the attention of naturalists.

Placerville, California, *Mountain Democrat*
November 10, 1877

Thirty-one Feet of Snake

T. O. Carter and Daniel Cleves, of Antioch, Cal., while riding
horseback in the Diablo Hills, near Round Valley, saw a monster
snake in a semi-dormant state a few feet ahead. Carter, who had
a shotgun, emptied both barrels at the head of the serpent. When
assured that life was extinct, Cleves measured the snake, and
found its length to be 31 feet. The body was from three to four
inches in diameter. The snake was of a greenish color, and had
apparently just shed his winter coat.—Antioch *Leader*.

New Philadelphia, Ohio, *Democrat*
May 16, 1878

The "Boss" Snake
The Monster of California, and a Narrative of its Discoverer.

[From the Calaveras *Chronicle*.]
On the 12th of August, 1868, the serpent was first seen in the
vicinity of Zane's ranch, near Spring valley. Several persons—
reputable people—saw the monster on two or three occasions, but
always at a considerable distance—never nearer than a quarter
of a mile. The reptile created the most intense excitement in the
neighborhood, and at one time the getting up of a party to hunt it
down was strongly agitated. What were then thought to be the
most extravagant stories regarding the size of the serpent were
told, but recent events prove that the truth was not exaggerated.
The snake was seen in an open field in broad daylight, and

described as "being from forty to sixty feet long, and as large around as a barrel." The mark of the monster in the dust where it crossed the road bore witness to its immense proportions. There was a difference of opinion regarding its method of locomotion, some maintaining that it progressed by drawing itself into immense folds, after the manner of a caterpillar, while others were equally certain that its motion was similar to others of the ophidian family. The serpent disappeared for several months, and was seen by Mr. W. P. Peek, of this place, while coming up the hill from the Gwin mine. Mr. Peek was driving a two horse team and had got about half way up the steep hill that has to be ascended in leaving the mine, when he heard what he supposed to be the loud "screeching" noise sometimes made by a wagon brake. Certain that a team was coming down the grade, and being in a favorable place for passing, he turned out of the road. After waiting until out of patience, and no team appearing, he drove on. He had gone but a short distance when a movement in the dense chaparral that lined the road, attracted his attention, and, advancing in the direction, he was horrified by the sight of a portion of the body of an immense serpent. At the same time his horse became unmanageable, and while Mr. Peek's utmost endeavors were put forth to prevent the escape of the frightened team, the monarch moved slowly off into the brush, making the hissing sound he had mistaken for the brake of an approaching wagon. About a year subsequently the serpent was seen by a couple of boys in the vicinity of Mosquito, the youths being so badly frightened that they could scarcely reach home and tell the story. Such is briefly the story of the Calaveras serpent up to Saturday of last week, when the experiences had with it at once settled all doubts as to its reality, and fix the fact beyond question that one of the largest boas of which we have knowledge has its residence in this county.

A Frenchman named Raud makes following statement of his experience with a serpent, to which he and his partner, F. C. Buylick, express their willingness to make affidavit. They are both Frenchmen, and are engaged in cutting wood and burning charcoal. Mr. Raud had shot and wounded a hare, which he followed into a thicket, and thus tells the story:

"I had proceeded twenty-five yards, perhaps, when I emerged into an open space not to exceed thirty feet in diameter. As I entered it the hare dragged itself into the brush on the opposite

side, and I quickened my steps in pursuit. Almost at the same
instant I was startled by a loud, shrill, prolonged hiss, a sound
that closely resembled the escape of steam from the cylinder of a
locomotive when starting a heavy train. I stopped as suddenly as
if my progress had been arrested by a rifle bullet, and looking
toward the upper end of the plat my eyes encountered an object
the recollection of which even now makes me shiver with horror.
Coiled up not more than twenty feet from where I stood was an
immense serpent—the most hideously frightful monster that ever
confronted mortal man. It was a moment before my dazed senses
could comprehend the dreadful peril that threatened me. As the
truth of my terrible, situation dawned upon me, my first impulse
was to fly; but not a limb or muscle moved in obedience to the
effort of my will. I was as incapable of motion as if I had been
hewn in marble: I essayed to cry for help but the effort at articu-
lation died away in a gurgling sound upon my lips. The serpent
lay in three great coils, its head, and some ten feet of its body
projecting above, swaying to and fro in undulatory sinuous, wavy
convulsions, like the tentacles of an octopus in the swift current
of an ebbing tide. The monster stared at me with its great, hate-
ful, lidless eyes, ever and anon darting its head menacingly in my
direction, thrusting out its forked tongue, and emitting hisses so
vehemently that I felt its baleful breath upon my cheek. Arching
its neck the serpent would dilate its immense jaws until its head
would measure at least eighteen inches across, then dart toward
me, distending its mouth and exhibiting its great hooked fangs
that looked like the talons of a vulture. As I stood in momentary
expectation of feeling the tusks and being crushed in the con-
stricting fold of the scaly monster, my situation was appalling
beyond description—beyond the conception of the most vivid
imagination. The blood ran down my back cold as Greenland ice
and congealed in my veins. Every pulse in my body seemed to
stand still and my heart ceased to beat. Even respiration was slow
and painful. There was a choking, suffocating sensation in my
throat, and my lips became dry and parched. There was a ringing
in my ears, dark spots floated before my eyes, and I should have
fainted but for the horrifying reflection that if I gave way to such
weakness my doom was inevitable. A cold clammy perspiration
oozed from every poor, and so intense was my agony of fear that
I suffered the tortures of the damned augmented a thousand fold.

While all my physical capacities were prostrated and para-lyzed, every mental faculty seemed preternaturally sharpened. It appeared as if the terrible tension of my nerve and bodily inca-pacity immeasurably increased my range of vision, and rendered my perceptive faculties critically acute. Not the slightest move-ment of the serpent escaped me, and every detail of its appear-ance—size, color, shape and position—is, alas! only too strongly photographed upon my recollection. As I stated before, the serpent lay in three immense coils, the triple thicknesses of its body standing as high as my shoulders. The monster was fully twenty inches in diameter in the largest place. Its head was comparatively large. Its tremendous jaws that at times dilated to twice their natural size, having enormous hooked fangs that fitted in between each other when the mouth was shut. The neck was slender and tapering. The belly of the serpent was a dirty whitish color, deeply furrowed with transverse corrugations. With the exception of about ten feet of the neck and contiguous parts which were nearly black, the body of the snake was brown, beauti-fully mottled with orange-colored spots on the back. How long I confronted this terrible shape I do not know. Probably only a few moments; but to me it seemed ages. At length the serpent began slowly to uncoil, but whether for the purpose of attacking me or retreating I could not fathom. You can have but a faint concep-tion of my relief and joy when I discovered that it was the latter. Lowering its crest and giving vent to a venomous hiss, the monster went slowly crashing, through the chaparral, its head being plainly visible above the jungle. For a moment I could scarcely realize that I was no longer threatened by a death too horrible to con-template. There was a tingling sensation through my body from the top of my head to the soles of my feet as the blood again commenced circulating in my veins. I attempted to step forward, but so benumbed were my limbs that I fell heavily to the earth. Recovering, I staggered through the chaparral into the open country. As I emerged from the thicket I saw my partner a short distance up the ridge and motioned him to approach. When he did so he was greatly alarmed at my haggard appearance, and excitedly inquired the cause. In reply I pointed to the serpent, then some 100 yards distant—a sight that threw him into the utmost consternation. We watched the monster until it disap-peared from view in the rocky recesses of a cliff that overhangs the river. We were enabled to measure the length of the serpent very exactly by its passing parallel with two trees, its head being even with one while its tail reached the other. Mr. Buylick has since ascertained that the trees are forty feet apart."

Ottawa, Kansas, *Journal Triumph*
August 29, 1878

A report is current of a mammoth snake inhabiting the Sisquoc rancho, fifteen miles east of Santa Maria, Santa Barbara county, California. It is said that the drying up of the springs in the mountains has caused this monster reptile to seek lower lands for water. It is affirmed that its track measures five inches across, which would indicate that the snake would be from 15 to 20 inches in circumference.

Boise City, Idaho, *Idaho Tri-Weekly
Statesman*
September 5, 1882

California Serpents.
Several Accounts of Very Large Snakes in the Yuba Country.
From the Grass Valley Tidings.

Four years ago Fred Campbell was down in the neighborhood of Smartsville, to which place he was delivering a wagon load of soda water. We mention what the cargo was so that it may not be said that he had bad whiskey aboard. A man was with Campbell. As they were going along the road near Mooney's Flat they heard a noise in the bush fence on one side, and, looking, they saw the head of a huge serpent emerge from the fence. The snake was a monster in size, and Campbell and his companion looked closely at it. They described the body, about the middle, to be as large as a flour barrel, and the length of the snake was certainly over fifty-one feet. They ascertained this approximate length by observing that when the tail was just leaving the fence the head of the serpent was near a certain bush, and the bush was about forty-one feet from the hole in the fence where the snake broke through.

This distance was measured. but the snake was not in a straight line between the two point, for it had the undulations that all serpents use to enable them to crawl. The big snake moved very rapidly toward the Yuba river and was in sight for only some five seconds.

Before Campbell saw this monster, many persons had reported seeing a very large snake in the vicinity of Industry Bar, at the junction of the Yuba river and Deer creek. Since Fred Campbell's narrative, every once and awhile we have heard of persons seeing a great snake somewhere on the line of one of the branches of the Yuba. Near French Corral, Freeman's Crossing, and other places, this kind of a big snake has been reported as being seen. The latest account we have is from Dr. Holdsworth, an excellent and truthful gentleman, who resides at Milton, in Sierra county. He says that during the latest snow storm he saw a track made in the snow as if a snow shoe had passed along, but the shape of the track and the route that it followed showed that no human being could have made it. About a week ago he found out what made the track, for then he saw sporting in the water of the creek near where the tracks were a huge monster, fully twenty feet long, which resembled a snake. We have no doubt but the snake Dr. Holdsworth saw is the younger brother or younger sister of the snake Fred Campbell saw near Smartsville. We seriously believe there is a family, so to speak; of gigantic indigenous serpents that inhabit the banks of the Yuba river in the foot hills and the higher mountain regions. Too many have seen such things to allow a doubt of their existence.

Syracuse, New York, *Standard*
July 7, 1884

Yuba River Serpents.
They Are Tremendous in Size and Fierce as Tigers.
Nevertheless Fred Campbell Managed to Beat One with a Siphon—
Sight Which Those Who Beheld It Will Never Forget.

"There isn't the least bit of doubt," said Ranchman George Wilmot, of Grass Valley, Cal., to a New York *Sun* correspondent, "that there is a race, or at least a family, of monster snakes indigenous to the foothills of the mountains that rise above the Yuba River valley. Some tremendous serpents have been seen there. The biggest one of these that I ever heard of being seen was seen by a man named Fred Campbell, who drives, or used to drive, a soda-water wagon, delivering bottled goods at different places along the river. He and a man named Collamer were driving along the Smartsville road one day, and when near Mooney's flat they heard a noise in a brush fence at one side of the road.

"'Great Butter's ghost!' exclaimed Campbell. 'Look there, Collamer!'

"He didn't have to tell Collamer to look, for Collamer was looking, and looking so hard that his eyes were hanging out on his cheeks, as well they might, for, sticking out of the brush, not twenty feet away, was a snake's head so big that Campbell declares it couldn't have been forced into a six-gallon beer keg. The snake the head belonged to soon began to make its presence known, and Campbell could trace the space it covered by the swaying of the bushes, which swaying, as if a hard wind were agitating the bushes, extended back for more than fifty feet. The great serpent had its big green eyes fixed on Campbell, who says that the tongue that shot in and out of the snake's mouth looked like a foot and a half carving fork painted red. The snake plainly was bent on taking in Campbell or Collamer or the horse and wagon, or all of them, for it came right on toward them out of the brush, advancing slowly but steadily. The horse had stopped, and stood in the road trembling as if paralyzed with fear. All this time Collamer hadn't said a word, but kept his bulging eyes fixed on the snake, just as if he had been charmed by it. Campbell had no weapon of any kind, but he was a resourceful chap, and not much given to sitting down and letting things get away with him without making an effort to prevent it. While that huge Yuba river serpent was drawing toward him, Campbell got an idea. He didn't know whether it would work or not, but he had the nerve to try it. The snake's big head was at last within a yard of the wagon, and was raised even with the seat. Campbell quietly picked up a siphon of carbonic water and, taking good aim, pressed the valve and shot a swift stream into one of the serpent's eyes, and before the snake lowered its head filled the other eye with the water

"Campbell says it was a sight to see that snake when it got the stream in its eyes. It dropped its head on the ground, thrashed it around, and hissed like escaping steam from an engine, then it reared again, and thrust its head out savagely toward Campbell, who shot another stream from the siphon into both its eyes. That was enough for the serpent, enormous as it was, and Campbell says it went squirming and twisting away.

"'He went pretty fast,' Campbell says, 'but he was at least half a minute passing a given point, and the smallest part of his body, except his tail, was as big as a pine log!'

"The snake went straight for the Yuba river, plunged in, and made the water fly as it swam down the stream as far as Campbell could see. It had been out of sight more than a minute, Campbell says, before Collamer's eyes began to go back into their sockets again and it was five minutes before Collamer could say a word, he was so far gone with terror. Then all he said was:

'If anybody ever tells me now that he saw a jack rabbit once that weighed a ton, I'll believe him!'

"This great serpent, or its twin brother or sister, has since been seen at Industry Bar, French Corral, Milton and other places, and Doc Holdridge, of Sierra county, says that he saw another member of this family of gigantic Yuba river snakes in that county, but he doesn't think it was more than thirty feet long—a young fellow, probably. It was big enough and old enough, though, to catch a deer and drag it away into the chapparal, right under the doctor's nose. Doc was worked up so that he forgot he had his Winchester with him until the snake was out of sight with the deer.

"'I might just as well have pounded the snake full of lead and anchored him there as not,' Doc says, 'and I ought to be shot myself for losing my head.'

"I haven't seen any of these big snakes yet myself, but I'd like to. I think that when I go back I'll gun for one of 'em. It seems to me that I'd rather have the skin of one of those snakes than half a dozen grizzly pelts."

Decatur, Illinois, *Daily Republican*
July 19, 1895

Latest Thing in Snake Stories
Sonoma County Women See a Boa That is Two Feet in Diameter.

Santa Rosa, Sept. 1. A boa constrictor is at liberty in the hills of Alexander valley and an organized hunt will shortly be made to put an end to its existence. It has frequently been seen and more frequently signs of its presence have been found. Yesterday the mammoth snake was stretched alongside a fence on the county road and caused a horse to take fright and run away with the occupants of a buggy, Mrs. E. B. W—, Mrs. William Powell of Eureka, Master Ralph Powell and Miss Katie Powell of Alexander valley. These ladies were driving along the road and believed the object to be a huge log when they first came in view. The horse they were driving suddenly balked, but under the urging of the whip passed the snake with a wild snort. The serpent is described by the ladies as fully twenty foot in length, with a head as large as that of an ordinary child of 12 years. Its body was more than two feet in diameter.

Joseph Alexander has known of the presence of this mammoth reptile on the ranch for thirty years and he is organizing a body of men to hunt and slay it. The ranch is directly on the road between Cloverdale and Calistoga, over which the circuses traveled thirty years ago in wagons. It is surmised that as a tiny snake the reptile got away from one of the circus wagons and has continued to live and grow on the ranch. Its presence is very undesirable and every effort will be made to rid the ranch of the reptile.

Fairbanks, Alaska, Evening News
September 1, 1906

Big Snake Seen by Indians Only

The Indians on the reservation a few miles from this city have become greatly alarmed of late over the appearance of a large snake on their grounds, which is described by them as being as large as an ordinary stovepipe and about ten feet in length. They refuse to leave their homes unless accompanied by a white man, as they think the great monster has come to take their lives and they will not venture from their huts, where a large fire is kept day and night in an effort to scare the intruder away.

The Indians, being too greatly alarmed to kill the reptile, sought the aid of white men, but upon their appearance the snake could not be found.

At night the reds claim that they can hear the snake crawling around their huts and the hissing sounds made by it are distinctly heard. They also say that this is the same snake that was seen in Alexander valley last year by a tribe of Indians on the reservation and which has been inhabiting northern Sonoma and southern Mendocino counties for several years.
—Ukiah Correspondent, San Francisco Cali.

Sheboygan, Wisconsin, *Daily Press*
March 23, 1911

Posse at Decoto, Armed to Teeth, Chasing Ten-Foot Boa Constrictor
Huge Snake, Which Escaped From Circus as Baby, Again Reported Seen Near Town.

Decoto. June 10.—A number of-years ago a circus stopped at Hayward leaving behind rumors of an escaped baby boa-constrictor some four feet long.

Every year someone within a ten mile radius of Hayward has seen that rumored snake—usually on Saturday night or Sunday morning—and every time it was seen it had grown a foot.

This morning it was seen again with 10 feet of tail stretching behind its head and a baleful look in its eye. As a result the entire town of Decoto is out snake hunting on the Henry May ranch one-half mile outside the city limits, for the man who reported it is an undisputed teetotaller.

Juan Costa, who discovered it today, is a gardener on the May estate. He swears that he stumbled over the reptile and that it was big enough to be a tree, only it moved.

About 9 o'clock he came running into town completely exhausted asking for help to capture or kill the giant snake. He stated that while he was cleaning brush in the bed of. Dry Creek, a gully which forms a center for a picnic ground maintained by the May family, he heard a rustling noise and right before him a black, green and yellow snake fully 8 inches in diameter slid across his path and disappeared.

Mrs. May, left on the ranch, telephoned to Chief Walter Walker of the volunteer fire department to come out at once, and he organized a snake hunting party armed with lassos, shotguns and monkey-wrenches and departed for the snake's alleged hiding place.

Oakland, California, *Tribune*
June 10, 1926

Sea Serpent, 25 Feet Long, Caught In Mine Shaft

Martinez, Cal., Nov. 14—(UP)—Excitement ran high here today over the great sea-serpent hunt, a hunt said to have ended in the early hours of this morning when the "sea-serpent of Old River" allegedly was carried away in a popular priced motor car.

The stories were substantiated by all the best citizens and in fact by everything except production of the serpent.

The monster, according to the generally circulated story, was trapped in an abandoned mine on the slopes of Mt. Diablo near the old ghost town of Somersville and there cried itself into exhaustion beneath a withering attack from tear gas.

Thus broken down with grief, the serpent allegedly was lassoed, poured into the automobile and carted away—maybe to a zoo or as a few skeptics hinted, to resume its peacetime duties as a section of fire hose.

Sea-serpent stories have been emanating from the delta area of the San Joaquin for many weeks. Numerous persons reportedly had seen the monster disporting itself in the tule swamps.

Mrs. Fred Seamm, wife of an Antioch city councilman, and Mrs. Alice Walbi, a friend, started a widespread search for the snake when they reported having seen it while driving in the vicinity of Antioch.

Two men, it seems, were retained by the Antioch city council some time ago to capture the serpent. It was agreed that the fee for the creature, dead or alive, would be $50.

The search centered in the Somersville area. About 2 a. m. today a restaurant proprietor drove into the country with sandwiches and coffee for the hunters.

He reported the mysterious creature had been cornered in the coal mine, located near the foot of Mt. Diablo. Tear gas had been discharged into the tunnel. Mack became overanxious and headed into the shaft, but was forced back when he nearly was overcome by the fumes.

Meantime, a crowd gathered. The beast was said to have been lassoed with a rope and apparently injured when it was withdrawn.

Witnesses said the snake was placed in an automobile and taken away soon after it had been captured. They described it as about 25 feet long and from six to eight inches in diameter.

Descriptions checked closely with those of others who claimed to have seen the serpent on a dozen or more occasions, at various points in the San Joaquin river bottoms.

Efforts were being made to locate the corpus delicti at zoos and other places in the Oakland and San Francisco districts.

Reno, Nevada, *Nevada State Journal*
November 15, 1934

Colorado

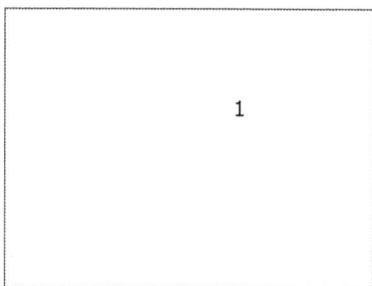

1. Castle Rock, Douglas County

Hunt Huge Boa Constrictor

Castle Rock, Colo., March 20.—(U.P.)—Ranchers claimed today that a giant South American boa constrictor which has survived three of Colorado's sub-zero winters was loose in this high-altitude cattle-raising area. They said the huge snake, which kills by crushing its prey in its coils, left a track four inches wide, a tread which an authority said would indicate that the constrictor was about 19 feet in length.

Hartford, Connecticut, *Courant*
March 21, 1950

Connecticut

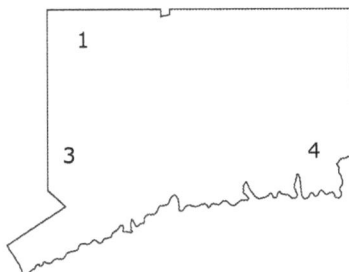

1. Canaan, Litchfield County
2. Housatonic River, not denoted (multiple western counties, north into Massachusetts)
3. Lake Kenosia, Fairfield County
4. Hamburg, New London County

A Snake Sensation in Connecticut—
Consternation among the People.

They have a first-class sensation in Canaan, Conn., in the shape of a big snake. It has been seen every morning for the last fifteen years in nearly the same spot. It was seen but a short time ago by a gentleman of the highest respectability, who is well known, and no one would think of doubting his statements. He was passing along the highway where the snake has been generally seen, not thinking anything about it, when suddenly his eye caught something lying under a large elm tree, which looked like the shadow of a large limb. Its shining brilliancy, and the real shadow of the tree being opposite from where the object lay, the thought instantly entered his mind that it was nothing else than the big snake he had heard so much about, and so it proved to be, for it immediately started for the swamp. The animal having to cross a road he had a fair view of his snakeship. A thrill of horror passed through his veins as the monster crushed its way through the brush with almost the speed of lightning into the swamp, and was immediately lost to sight. He thinks it would measure thirty inches around its body in the largest place; in fact it appeared nearly to be one size, except three feet from the end of the tail, which tapered considerably toward the end. Its length he should judge to be not far from twenty-eight to thirty feet. Its skin was so black and bright that it fairly dazzled. The farmers in the vicinity complain less of the depredations of the reptile than formerly.

On the border of the swamp where it lives graze a large herd of cows, and it is among these it gets its living. The owners of these cows have for some time past wondered why some of their best milkers had failed to give their usual quantity of milk; as it is only in the morning the owners miss it, they have come to the conclusion that the snake sucks them, and thus gets its living. It is well known that snakes are very fond of warm new milk, and it is this kind of food no doubt that has caused this one to grow to such immense proportions.

There is a reward of one hundred dollars offered by private citizens for its capture, and fifty if killed. There is a strong pressure bearing upon the town authorities to offer a reward large enough to secure its capture or destruction, and the subject will be brought before the annual town meeting the first Monday in October. Hundreds who attended the camp meeting lately held there came more for the purpose of seeing this monster snake caught than to seek for the straight and narrow way. Few could be induced to stay on the grounds nights for fear it would make them a visit. It was this one thing more than any other that made the meeting almost a failure. One man has determined to sell his farm, and leave the place. His wife has already left and says she will never live on the place again until the snake is killed. There

is a perfect panic among the people, and what will be the result
time alone can only determine.

Janesville, Wisconsin, *Gazette*

October 7, 1867

A Big Snake Story.

Some seven or eight years ago, says a Poughkeepsie (N. Y.)
letter to the New York *World*, much alarm was manifested
along the line of the Housatonic Railroad by the appearance
of a tremendous snake, but of what ophidian species he was a
sample no one could tell. The reptile made its home in a dense
swamp in the vicinity of the railroad track. Time and again the
reptile was shot at by hunting parties, but it always managed
to got away safely. The track of the Housatonic Railroad runs
directly through the swamp named. Early one morning the engineer of a passenger train suddenly discovered a long black object
lying across the rails and he shut off steam and whistled for
brakes. While the speed was slackening the engineer saw that it
was the veritable snake, about which so much had been said, and
he pulled the throttle again, intending, if possible, to run it down
and cut it in two, but just before the engine reach it the end of
the tail slid off the rail and almost immediately the entire snake
disappeared. Four or five years ago two men riding a buggy along
a road which skirts the swamp say they saw the now famous snake
wriggling slowly across the road. They were certain, from the
measurement of the ground where they first saw it, that it was
over twenty feet long. Again the snake excitement broke out in
the neighborhood, and parties were organized to hunt the reptile
down, without success, and the interest in the matter soon died
out. The cause of the disappearance of the snake for the last four
years has been a great mystery, but it has been completely solved
in an extraordinary manner. Some imagined that the snake has
found his way to the Housatonic River, and, escaping death by
drowning, thence to the ocean, and became a veritable sea serpent.
Others held that he had taken to the mountains. Really he never
left the swamp of his birth, for his colossal remains have been
discovered. On Saturday last two men named Kelly and Smith,

both well and favorably known in the neighborhood, went into the swamp with a sled to get a load of wood. After a while they same to a large buttonwood tree which had fallen to the ground. They discovered it was hollow, and in order to handle it easily they attempted to saw it up. They had sawed nearly through one part when suddenly the saw grated as though it had struck a stone. Thinking that it was a curious place to find a stone they at once went to work to investigate. They plied their axes and by dint of hard labor split the tree, when to their astonishment the obstacles which the saw had struck proved to be bones. Then they opened the tree as far as they could find bones, some 21 feet, and the remains proved to be those of the monstrous reptile so often seen but never captured. The ribs measured six inches in diameter and from that tapered down to smaller sizes. The tail of the snake was found imbedded in the upper part of the tree, and both men believe he went in backward. It is believed that the last time he was chased, some four years ago, he sought refuge in this tree and never came out again. The bones and other remains have been preserved and will be sent to the Society of Natural Science, together with a history of the case.

Decatur, Illinois, *Review*
March 14, 1879

A Fight With a Very Large Snake.

The following snake story is told in a recent Norwich (Conn.) telegram: The biggest snake Connecticut has produced in this century was killed in Westford, Windham County, one day last week. Mr. Allen, who killed the snake, is an enterprising farmer of unimpeachable veracity, and his story of the fight is corroborated by a coterie of friends who saw and assisted in the scene.

Mr. Allen had been trimming his orchard trees, using a keen-edged sickle to lop off the sprouts. He was interrupted by a party of acquaintances, who drove to his door and asked him to guide them to a huckleberry pasture on his farm. He cheerfully consented, and started off at the head of the troop in his shirt sleeves and with his sickle in his hand. A monster boulder uplifts itself at the entrance to the pasture, and is girt on one side by a narrow ledge that seems to have been made especially for snakes to sun themselves on. As Mr. Allen passed this rock he saw an enormous black coil as big as a rubber tire hose lying motionless on the stone. It was a great serpent, fast asleep, basking in the sun. A young and incautious member of the party at once hurled his dinner-pail at the snake, striking it in the center of the coil. The snake sleepily raised its head, saw the empty dinner-pail and the party gazing with deepening interest.

Mr. Allen avers, and his friends also say, that the serpent eyed them for an instant with a glittering, icy smile, its tail vibrating with such rapidity that it was invisible at first. Its tongue darted in and out of its jaws with almost inconceivable velocity. The first impulse of the party was to turn and run. They plunged through the blackberry and huckleberry bushes in headlong panic, Mr. Allen covering the rear. Looking over his shoulders, he saw the snake following, with its mouth wide open. He continued his flight, but the serpent gained on him, and at length, just as the serpent's head was overtopping his shoulder, he turned suddenly and cut at the serpent's crest and neck with the keen-bladed sickle. The snake dodged the blows, and the party, ceasing their flight, one by one closed in on the snake with fence-rails, blackberry briers, stone slabs, and other missiles, and there was a long and fierce fight. It was not until a quarter of an acre of pasture had been tramped over that the snake was killed. It measured just eleven feet and five inches, and the thickest part of its body was as large around as an ordinary teakettle. In its stomach were found the remains of Mr. Allen's choicest rooster, which he had missed from his roost a few days previously. As many other fowls had mysteriously disappeared lately, Mr. Allen conjectures that they had been seized by the snake.

St. Joseph, Michigan, *Herald*
September 27, 1879

Lake Kenosia's Monster Caught

Danbury, Conn., Dec. 8.—The strange monster which has frequently been seen in Lake Kenosia has been captured. It is a monster serpent, of an unknown species and its immense size was not exaggerated by those who saw it at various times. Warren

C. Baker, a charcoal burner who has a pit near the lake, was driving along the shore last evening, when he found the serpent, lying dormant and nearly frozen on the sand. Its immense size frightened him, and he drove to the hotel for aid. The serpent was securely bound with ropes and brought to this city, where hundreds have seen it. Its length is 19 feet 8 inches, and its body is 33 inches in circumference. Its head is flat and its body is covered with scales of a black and brown color.

Indiana, Pennsylvania, *County Gazette*
December 10, 1890

Adventure With a Snake.
It Results in the Breaking of Jerry Canfield's Two Legs.

Danbury, Conn., Aug. 25.—Jerry Canfield of Hamburg is suffering with two badly fractured legs, the result of a thrilling experience with a monster snake. Canfield was sawing logs at a mill in Indian Woods Thursday afternoon, when he was startled by a rabbit that darted from the bushes and sprang over the log between him and the saw. Behind the rabbit glided a large black snake of the white-throated species. As the snake crawled over the log the carriage bore it to the saw, which cut off a piece of the tail. Angered by the pain the snake darted in the direction of Canfield and in an instant was coiled tightly about his legs.

The Snake was a Monster.
Canfield attempted to spring backward, but the serpent had bound his legs together and he fell between the braces and stringers into the carriage pit below. Somehow the unfortunate man's legs were drawn under the carriage and the trundles or iron wheels that support it passed over his legs, breaking them like pipestems and at the same time killing the snake. It was one of the largest snakes of the kind ever seen in Connecticut, measuring 11 ft 3 in. in length. Canfield was removed to the Hartford hospital, but it is not certain that he will recover.

Decatur, Illinois, *Morning Review*
August 26, 1891

Delaware

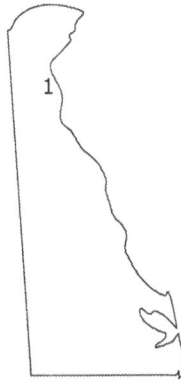

1. Delaware City, New Castle County

A Delaware Snake.

Delaware City has another sensation, and this time it is one that is positively startling if the story told is true. There can be no doubt, however, of its correctness, inasmuch as it is vouched for by James Cheeseman, said to be "a gentleman beyond reproach and of unimpeachable character." That gentleman states that while driving leisurely along the road from Delaware City to St. George's, on Tuesday morning last, he was suddenly startled, while on the St. George's causeway near Dragon Creek, by the

appearance of a large reptile coming down the road directly toward him at a rapid pace. As the moving mass approached he
discovered it was a huge black snake at least twenty feet long and
about a foot in diameter, as near as could be judged by the
hurried glance he gave the "varmint."

Before the loathsome creature reached him he gave his horse
a cut, with a whip just as the snake made a jump for the animal.
The horse sprung to one side and started on a dead run, apparently terribly frightened. The snake missed its aim, but struck
the front wheel of the carriage a stunning blow, breaking out
nearly every spoke and making the vehicle tremble and cant
dangerously to one side. Mr. Cheeseman did not succeed in stopping his horse until it had run about a mile, when he looked back,
but saw nothing of the horrible reptile, which he describes as
being covered with large scales.

Charles Brown and his wife, while blackberrying in the same
vicinity last week, were chased by the same immense creature, but
by dint of hard running they succeeded in gaining a place of shelter.

The existence of the snake is well known, and a gentleman
who does not care, from modesty, to have his name appear, very
reluctantly gives an experience he passed through about ten days
ago. He started out blackberrying in the marshes along the bank
of Dragon Creek, carrying with him also a shotgun. Feeling tired
about noon he concluded to rest awhile and began to make his
way to an old log, as he thought, about twenty feet distant, to
take a seat. Upon approaching what he believed to be a fallen
tree, the object moved slightly, when, he discovered it was not a
log, but an immense snake, on which were scales about the size
of soup plates. Instinctively he fled as rapidly as the nature of the
ground would permit, leaving his gun behind. Finding he was not
pursued he glanced back and saw his snakeship moving slowly away,
and although he believes he saw the middle of the creature first, it
was fully five minutes before the tail passed through the bushes.

A crackling noise followed in the wake of the snake, and small
trees shook as the creature's tail swayed from side to side. Gaining
courage, the *Gazette* informant secured his gun and started after
the snake, which had disappeared in the woods. After traveling
about a mile the hunter came in sight of the trunk of an old hollow
oak, about fifty feet high and two feet in diameter, from the top
of which protruded the head and almost ten feet of the "awfullest

thing I ever saw," as our informant remarked. Its head was about the size of a nail-keg and of a glossy black color, its forked tongue shooting out venomously fully a foot, accompanied by a hissing sound resembling escaping steam from a locomotive. The hunter took aim and fired, the load of shot striking the snake, the shooter thinks, two feet from the head. The lead failed lo make any impression, but dropped to the ground flattened out. A shot from the second barrel resulted the same way. The last report, however, seemed to anger the snake, and it started out of the top of the tree, its fire-like eyes gleaming in the sunlight. The man at once left, looking around when some distance off and seeing the loathsome creature slowly letting itself down by the tail to the ground, while around the top of the tree appeared the heads of numerous small snakes—perhaps a hundred—evidently the young of the monster.

The now thoroughly-frightened berry-gatherer fled precipitately, and not until Mr. Cheeseman's experience became known did he tell of his adventure, which of course will have to be taken *cum grano salts.*

The existence of a huge snake in the vicinity mentioned has been known for some years, and it is on record that Henry and William Carson, sons of John Carson, were at one time chased home while berrying. At that time a party was organized, and, armed with guns, pistols, hatchets, pitchforks, etc., scoured the vicinity for two days, but without success. Another party is to be formed in a few days, and a grand snake hunt organized.— Wilmington (Del.) *Gazette.*

Decatur, Illinois, *Weekly Republican*
November 9, 1882

Florida

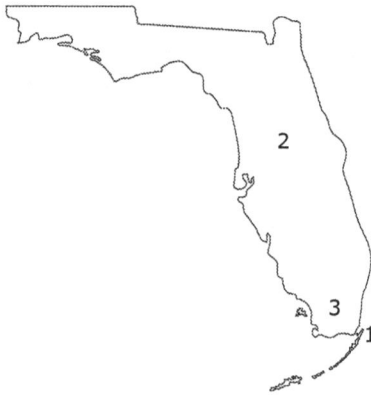

1. Black Point Key, Biscayne Bay
2. Mascotte, Lake County
3. Everglades, extends over the southern region of the Florida peninsula.

Besides the stories that follow, there is a report from 1962 of a 30-foot long "sea serpent" sighted in and around the Myakka River. One herpetologist apparently suggested it could be an anaconda, but whatever it was, it disappeared without positive identification. An oarfish was later found beached on Manasota Key Beach, leading to speculation that it was responsible, but that seems unlikely. For now, I don't have enough details to identify this report as an actual big snake sighting. (Charlotte Harbor, FL, *Sun Herald*, January 14, 2006)

Snake Controls an Island.
Black Point Key, on the Florida Coast, Held by a Boa.

Black Point Key, a few miles below Miami, Fla., is now known as "Snake"' key, from the fact that a huge Brazilian boa has control of the island, much to the terror of the guides who go there and the few inhabitants of the close-by islands.

The story is that several years ago a steamer with a circus on board foundered off the island. The snake was then possibly ten feet long; now reliable persons state it is 30 feet long and of generous proportions otherwise. Indian guides have been afraid to go to the island since last fall, when two of their number were killed by some mysterious thing, and the snake was blamed for it. Several other persons who ventured to land have disappeared, and tourists now cannot get guides to go there to explore the island, which is about 50 acres in extent and thickly grown with low shrubs.

Walter Ralston, an agent of the Smithsonian Institution "snake department," has undertaken for the East Coast Railway company to capture the serpent, and he intends going to the island. He says that he has no fear of his ability to catch it. He has reliable reports of its existence, and thinks it is there, and doubtless a monster, too. He says he will prepare a huge canvas four by seven feet and will manage with bait to attract the serpent. When once it begins to crawl in it will continue going.

"After I once get him headed for the bag he's my meat," said Ralston, confidently. Ralston is thoroughly expert with serpents, handling all kinds with utter fearlessness.

San Antonio, Texas, *Daily Light*,
March 23, 1897

Chloroformed a Python.
Huge Monster on Snake Key Captured for Smithsonian Institution.

Walter Ralston, the snake agent of the Smithsonian institution, who went to Snake Key to capture a big python that had been the terror of the place, has returned to Miami, Fla., with the monster. It measures 34 1/2 feet long and weighs over 100 pounds. Ralston had a rough time capturing it. His party came upon it at the key early in the morning, while it was feeding upon a deer just caught. The reptile was furiously angry at being interrupted at its feast and hissed and thrashed around in a great manner. Ralston's two Indians worked hard to secure it, but they missed getting the snake's head in the long canvas bag provided for it, and the reptile's tail pulled around Ralston, hugging him up to a pine tree in a jiffy. His faithful Indians, though scared to death, manfully attacked the serpent, though Ralston would not let them hit it with an ax or cut it with their knives.

They pulled and pulled, but its folds became tighter around Ralston and he began to get frightened. Seizing a bottle of chloroform from his pocket, he threw it to the Indians, and they managed to open it and give the serpent, whose head had meanwhile been secured inside a canvas, a good dose. Shortly its folds related a little and Ralston struggled out free, but with aching ribs. The snake was then seized and tumbled into the sack. Ralston says that this is the small one, as there is a much larger one on the key that he will get later.

Decatur, Illinois, *Daily Review*
April 24, 1897

A Big Snake in Florida.
Fifteen Feet Long, with a Body as Big as a Six-Inch Stovepipe.

Mascotte, Fla., Aug. 20.—Mascotte has a snake story which is vouched for by several reliable witnesses. In Sunset Lake, directly south of town, a monster snake has been observed by a number of persons within the last four weeks.

Boys have come home from bathing with stories relating how they were chased by the great serpent.

All who have seen the snake unite in saying that it is at least fifteen feet in length, with an enormous head and mouth, large glittering eyes, and body as big as a "six-inch stovepipe." One person shot at it, but apparently did it no harm.

Several hunters are on the alert, anxious to win local fame by either capturing or killing this monster of the lake.

New York, New York, *Times*
August 21, 1897

Strange Serpent Slain in Everglades
"Dragon of the Everglades," from a Sketch by an Indian.

An enormous reptile, more like the mythical dragon than a land serpent, has been killed by a hunter in the lower Everglades. For 100 years it has not only been a tradition among the Seminole Indians, who live in the Florida everglades, than an immense serpent made its home in that region, and they affirm that two Indians had been carried off by the monster.

Recently Buster Ferrel, one of the boldest and most noted hunters in Okochobee, who for twenty years has made the border of the lake and the everglades his home, on one of his periodical expeditions into one of these lonesome wilds noted what he supposed to be the pathway of an immense alligator. For several days he visited the locality for the purpose of killing the saurian, but was unsuccessful in finding him.

Finally he decided to take a stand in a large cypress tree and await the coming of the alligator, taking provisions to last him several days.

For two days he stood on watch, with his rifle ready, but without the desired success. He was becoming discouraged, but determined to give one more day to the effort. On the third day, before he had been on his perch an hour, he was almost paralyzed by what looked to him to be an immense serpent gliding along the supposed alligator path. He estimated it to be anywhere from twenty to thirty feet long and fully ten to twelve inches in diameter where the head joined the body and as large around as a barrel ten feet farther back. The snake stopped within easy reach of his gun and raised its head to take a precautionary view of its surroundings. As it did Ferrel opened fire on it, shooting at its head. Taken by surprise, the serpent dashed into the marsh at railroad speed, while Ferrel kept up fire on it until he had emptied the magazine of his rifle, but failed to stop it.

About four days afterward he ventured back into the neighborhood to see how things were, and about a mile from where he first saw the snake he saw a large flock of buzzards and went to see what they were after, and there he found the creature dead and its body so badly torn by the buzzards that it was impossible to save the skin. He however secured its head and has it now in his home on the Kissimmee river. It is truly a frightful looking object, fully ten inches from jaw to jaw, with ugly, razorlike teeth.

Grand Rapids, WI, *Tribune*
January 25, 1902

From Willoughby's (1910) *Across the Everglades*:

With an apology for my slight digression, we will return to our coasting trip and our unsuccessful crocodile hunt on Card Sound. I had originally intended to hunt crocodile for a few days more, but, knowing the water in the Everglades would be getting lower as the month advanced, I considered it prudent to push south, knowing well where the largest crocodile lived, intending to return in March for a systematic hunt after my winter's work in the Everglades had been accomplished.

On the 31st of December we got the "Cupid" under way and ran as far as the mouth of Jew-Fish Creek. The creek is nothing but a very narrow passage which connects Barnes Sound with Black Water Sound. I am using here the names as given by the Geodetic Coast Survey Chart. Barnes Sound is known to the natives as Little Card Sound, and the northeast end of Florida Bay is called by them Barnes Sound. Comparatively little is known of these sounds, especially the northwest shores, as the few people who live in Key Largo and the smaller keys use the better water communication of the Hawk Channel for the transportation of their produce. Jew-Fish Creek has rather a blind entrance, which passes through a heavy growth of mangroves.

The water is quite deep and the current runs through with great swiftness, and is a fine place for all kinds of fish. The channel divides into two or three branches, the one running nearest Key Largo being the best. With a light wind we sailed across Black-water Sound, which is a pretty slice of water nearly circular, and about four miles in diameter, making a course nearly west, in order to find the opening which leads into the Bay of Florida, called Boggy Creek, a rather difficult place to get a sail-boat through, as the bottom is so soft that a pole pushes in to its full length.

There is a second opening to the bay, more to the northwest, which runs into several small land-locked bays. After much

difficulty we succeeded in getting through "Boggy," and anchored
for the night near its opening into the Bay of Florida. From here
I took a westerly course, which was somewhat out of my way, in
order to reconnoitre the uninhabited northwest shore of this large
bay, and also to prove the accuracy of a statement by Ed Brewer,
that a year and a half ago, while hunting on a certain creek, he
had killed a great snake of a kind that he had never seen before.
This creek leads directly to the Everglades. Brewer believed that
he could take me to the very spot, and thought that perhaps I
might find others of the same variety. I have for a long time
believed that in the southern part of Florida there exists a snake
of immense size. What this snake is has been my longing desire
to discover.

Many absurd and sensational stories have been circulating
this winter, without any truth in them whatsoever, concerning
the capture of this huge reptile. A circus charlatan brought a
python from the North in a box. He disappeared for a few days,
and returned to the settlement with a twelve-pound snake in a
bag, that he claimed to have captured on Long Key, which is in
the Everglades. I am quite sure that without an Indian guide this
man never reached that spot alone. But the fraud was quite
successful, and the newspapers were completely taken in. Taking
the sloop along the northwest shore as far as the water would
permit, Brewer and I left the "Cupid" in charge of Sam, and with
the "Coscochee" started for the "Land of the Big Snake."

Leaving the main part of the bay, we entered a smaller one,
which we crossed, and with unerring accuracy Brewer steered for
what appeared to be a solid wall of trees; but on reaching it a
hole in the foliage was seen, from which a fresh-water stream was
issuing. The water was perfectly clear, and just after entering I
could see many fish. Reaching for my light spear, I soon landed
three fresh-water garfish. Continuing further up the stream, the
foliage became more tangled and made our progress very slow,
until we reached what seemed to be the head of canoe navigation. The scene would rival some of those depicted on the Amazon
River. The hanging vines, the tropical foliage, obscured nearly
all light, and the air had an earthy, snaky smell.

Brewer, who had been pulling the canoe along by the over-
hanging branches, stopped and said, "We are near the place where
I killed the snake." I immediately felt for my revolver, thinking

there might be a whole den of pythons somewhere in the vicinity that had come to the funeral of their progenitor. The place itself was certainly the snakiest one I had ever been in. Brewer crawled on shore, with his head to the ground, and began lifting up the rotten leaves, examining everything very carefully.

This search went on for half an hour, when he said, "I have got it, but the buzzards have scattered the bones pretty bad!" I jumped out of the canoe at once and joined him, all eagerness to find a skeleton that might have belonged to the strange variety I have been in search of. Under the first layer of leaves was a section of vertebra, evidently that of a snake; its diameter, as compared with snakes I had usually seen, seemed very large. By clearing a space of ten feet around, we uncovered more pieces of the backbone and a great number of ribs. I returned to the canoe for a pail, and after a long period of careful work we recovered more than two-thirds of the entire body, but nothing as yet of the head, that I was so anxious to find; among the last two pieces, however, I discovered a bit of the jaw, with a large fang sticking in it.

That fang told me much, as I could plainly see the capillary tube running through it, which indicated to me rattlesnake. I did not, however, wish to decide hurriedly as to this fact, and preferred to wait until I had assembled the bones into the full structure, at the University of Pennsylvania, in consultation with the professors skilled in osteology. I got Brewer to tell me again of the killing of this reptile, a year and a half ago. It seems he had been hunting and exploring up this creek, in order to find a canoe passage that would lead to the Everglades. Pushing his way through the thick foliage, as we had done, he saw a little way ahead of him, on the low limb of a tree, what seemed to be the largest snake he had ever seen in his life. The snake he described as having longitudinal stripes, and not looking like a rattler. To pick up his shotgun and fire was the work of a second. The snake was badly riddled. He approached with the idea of dispatching him and taking his skin, when he was overpowered by a strong sickening smell, which caused him great sickness of the stomach and faintness. He lay in a partly unconscious condition for some time, and when he recovered sufficiently, he readied his canoe and made his way to the coast without paying any more attention to the snake. On my return to the North, the skeleton was carefully put together, and it was decided that it had belonged to a rattle-snake, but the formation of the vertebræ seemed to be different from those of the usual variety of the Florida rattler, which may account for the observation Brewer had made of the longitudinal stripes. In life it would have been about eight feet in length.

That great snakes of some species do exist in Florida, yet to be observed, I have not the slightest doubt. I pin my faith to the account that two different Indians have given me of snakes that were at least eighteen feet in length, and evidently belonged to the constrictor family. As I have remarked before, I have never found a Seminole to lie. I did not explore the head of this creek, but I am under the impression that it reaches to within two miles of the Everglades.

Floridians Report Big Snake

Miami, Fla., (U.P)—Two Miamians came in from the Everglades recently with a report they had seen a huge snake, at least 15 feet long, 10 inches in diameter, which looked like a log lying across the road. Experts believed it was a Gopher snake, but the discoverers temporarily called it "Glade Snake."

Hayward, California, *Daily Review*
July 13, 1931

Capture 'Grandbaby' of Boa That Escaped 53 Years Ago

Jacksonville, Fla.—(U.P.)—Two famed snake hunters, searching a dense swamp four miles from this city of 250,000 population for a 30-foot boa constrictor believed to have escaped from a circus train 53 years ago, captured its "grandchild" yesterday.

The monster reptile had been the subject of oft-repeated rumors that had grown into swamp legend for years.

When the search got into the open, anxious mothers ordered their children not to stray from cleared ground.

Ross Allen, head of a Florida reptile institute, and Dr. W. T. Neill, biologist and college professor, captured the 10-foot grandbaby with a body the size of a man's leg.

"Where this one came from, there probably are more," they said.

The 45-pound boa taken after a night-long hunt through the stagnant, vine-choked swamp, could easily crush the life from a man in minutes, Allen said.

It could swallow whole a small child or dog.

Children were ordered to stay close to their homes, and mothers peered often from windows to see if they obeyed.

The boa was found half-buried in an armadillo hole.

Instinctively timid, the snake lay still while Allen and Dr. Neill shoveled away dirt 10 feet away and uncovered the tail.

Allen pounced on the snake's head and Neill caught the other end.

"We stretched it," Allen said. "It's weak as a kitten when strung out."

Old timers along the swamp's rim remember when a circus train derailed there in 1897.

A giant boa constrictor was said to have escaped in the accident.

The snake story made the rounds for years but no responsible citizen would admit seeing a 30-foot reptile for fear of being branded a liar.

Then, two weeks ago, Cleveland Jones, retired engineer, came face to face with "it."

Johnson's hired man, M. M. Moore, also saw it, peering through bushes on the edge of Johnson's 100-acre tract.

"It's entirely possible the snake which escaped from the circus train gave birth later to several," Allen said. "They have 'em in litters like puppies."

Waterloo, Iowa, *Daily Courier*
May 21, 1950

Georgia

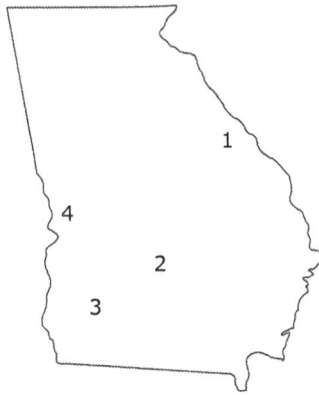

1. Augusta, Augusta-Richmond County
2. Wilcox County
3. Albany, Dougherty County
4. Columbus, Muscogee County

In Augusta, Ga., a water-snake eight feet two inches in length and ten inches in circumference, was recently killed by Alonzo Gilley.

Hagerstown, Maryland, *Herald and Torch Light,*
June 22, 1881

A Turkey Hunter's Adventure.
Hawkinsville *News.*

R. L. Duffie, while hunting turkeys in Wilcox County recently, had a strange adventure which is worth printing. He left home before daybreak, and just as it was light enough to see an object a few yards he entered the swamp where he knew the turkeys had a roosting-place. He seated himself at the root of a large tree, and in a few minutes he saw the drove of turkeys playing on the ground as they usually do when they leave their roost. Mr. Duffie laid his gun across a small log and began taking sight, when he noticed that the log either moved or his gun slipped. He readjusted the gun across the log, and again took aim, when the log moved again. This excited him, and he forgot the turkeys and began to wonder what could be the matter. On further investigation he was almost paralyzed with fear at finding that what he supposed was a log on which he had rested his gun was nothing more than a huge gopher snake. The reptile was about thirty-eight feet long, in the imagination of Duffie, and, although the gopher snake was never known to attack a man without being wounded or disturbed, Mr. Duffie thought it best to leave the swamp, and he did so. We shall probably hear something more from this monster reptile.

Marshfield, Wisconsin, *Times and Gazette*
December 1, 1883

Some five or six years since a monster snake was seen in Rawl's pond, six miles below Albany, and the gentleman who saw it related his adventure to a number of incredulous friends. It has remained to this day a matter of jest among the acquaintances of the frightened gentleman, the reported proportions of the huge reptile. Among the number who heard the statement with a mental reservation was Dr. C. P. Hartwell, who owns a plantation embracing the largest part of the swamp surrounding the pond. The doctor is now, however, a firm believer, but it required seeing to convince him. One day last week he was out in the field with his farm hands, and having occasion to go into the edge of the swamp, encountered a snake, the sight of which almost took his breath away. The reptile showed fight and the doctor called for his laborers. One responded, but seeing the snake, he fled. The others were called and with great poles the brigade surrounded his snakeship, and made on attack that resulted in his death. Dr. Hartwell says that he thought he had seen big snakes before, but all native snakes were midgets by this monster which was larger than any he ever saw in any circumstances. It proved to be of a variety of water snakes, about ten feet long, nearly as large as an ordinary man's thigh in circumference, with a very large head. When it opened its mouth it seemed possible for it to enclose a man's head.

Atlanta, Georgia, *Constitution*
July 7, 1887

Attacked By Coach Whip
Fulton Chokes Reptile

Columbus, GA, May 13—(Special)—It was only after a desperate fight that B. A. Fulton, superintendent of the Muscogee county poor farm, saved himself from suffocation in the coils of a monster coachwhip snake, 14 feet long and so big that he could not reach around him with one hand.

Mr. Fulton was walking over the farm several days ago and his little dog was trotting ahead of him. The dog spied the snake and began to bark, and the reptile came on to attack. The dog retreated and Mr. Fulton drew a hammer from his pocket and threw it at the head of the snake, missing it, however. The coachwhip, enraged, reared his head in the air and ran along the ground toward the gentleman with almost incredible rapidity. Before he could retreat the snake had reached him and began coiling about his legs, and in a few seconds time he was a helpless prisoner, his legs being bound as in a vise, while the angry reptile squeezing tighter with each fold was rapidly ascending his body. Just as the snake twisted around his body from the rear, its head being in front of his stomach, and as it was about to wrap another coil, Mr. Fulton seized it back of the head and began to choke the snake. He has a powerful grip and in this exciting situation, it was rendered still stronger by fear, and he soon choked the reptile so severely that he began to hastily uncoil. Seizing a favorable opportunity, Fulton stepped on the reptile's head with his heavy shoe and managed to crush its head, thus putting an end to one of the most savage coachwhips that has been seen in this county in a long time. For two or three days Mr. Fulton was almost prostrated from the shock of his strange encounter.

Atlanta, Georgia, *Constitution*
May 14, 1905

Illinois

1. Sandridge, Greene County
2. Sangamon County (including Lake Springfield)
3. Macoupin County
4. Warren County
5. Clay City, Clay County
6. Plymouth, Hancock County
7. Modoc, Randolph County

Snake Story.—The Carrollton *Democrat* says: "We are informed that Mr. Williams, who lives on what in known as the Sandridge, in this county, killed near his premises recently a monstrous black snake, measuring thirty-three feet in length. It was with great difficulty that he succeeded in killing it."

Alton, Illinois, *Weekly Telegraph*
May 25, 1866

A Big Snake

The Springfield *Journal* prints a communication from New City, Ill., dated 16th inst., which says: "The inhabitants of Cotton Hill Township, Sangamon County, residing near South Fork, have been considerably excited during the past week in regard to a monster serpent that has been seen near the bluff of South Fork, on land belonging to J. M. Haines, Esq., three miles south-west of Rochester. Mr. John Adwell states that on or about the 9th of August, while gathering berries near South Fork, he discovered a large snake or serpent about eighteen feet long. It was of a brown color, with yellowish spots on its sides, and apparently about twelve inches in circumference around the body. His wife saw the snake at the same time. He carefully measured the distance from one clump of bushes to another through which the snake was passing, and ascertained the same to be about fourteen feet, and he is satisfied that over four feet of the serpent had already passed out of sight. He afterward went back and saw it again near the place, but could not get a good view of it. Mr. L. Johnson states that three years ago he saw the same serpent or its mate near the same place. These statements are given in so honest and truthful a manner that your correspondent is fully convinced that there is a snake of unusual size in that vicinity."

New York, New York, *Times*
August 27, 1876

A Big Snake Story
The Monster Reptile That Swallowed a Three-year-old Child.

From the Rochester *Democrat.*

A tough story comes from Macoupin county, the truth of which we will not vouch for, but as our informant is a book agent, and claims that he witnessed it, it may have the same effect upon the minds of some that the truth could have, hence we give it. A lady named Smith, living near Carlinville, while engaged in washing clothes in her back yard last Monday morning, heard several screams from her little three-year-old child, which she had left asleep on a bed in the bedroom. She hastened to the room, and was horrified to see the child partially swallowed by a huge snake, only the little bare legs being visible, and the snake backing out of the front window with all possible haste. Mrs. Smith screamed for help, at the same time grabbing an ax and rushing to the rescue. She overtook the monster just before it reached the timber, and dealt it a blow with the keen blade of the ax just

A Few Inches Behind the Ears,
which disabled his snakeship from further locomotion. Our informant happened to pass just at the moment, and noticing the contest hastened to render his assistance by seizing the ax from the mother and cutting into the body, the mother meanwhile taking the child by the legs and trying to pull it from its prison, which she finally did by bracing her feet against the snake's under lip and taking a long, steady pull. The rapid blows of the book agent soon laid the monster lifeless, and the two turned their attention to the child, who was found to be uninjured beyond being slightly choked. The snake, when measured, was found to be eighteen feet long, and forty-four inches in circumference. When cut open its body was found to contain seven iron pots, two young calves, four ladies' bustles, nine guns, a sevenup deck of cards, three bushels of brass buttons, four pigs and a member of Congress certificate of election, besides a volume of Joaquin Miller's poems and a copy of the Carlinville Enquirer. Our informant says he has seen snakes in every quarter of the glebe, and even in his boots, but in size this one is the largest he has ever encountered.

Reno, Nevada, *Evening Gazette*
October 17, 1879

A Snake Sensation.

Though rather late in the season Ellison Township in Warren County comes to the front with a snake story that equals any sea-serpent sensation in Lake Michigan this year. One day recently Ed Davis of the town of Kirkwood was driving along a road that led by a stone quarry. Suddenly there appeared before him a monstrous snake, that frightened both his horses and himself. He declares that it was fully fifty feet long, and as it glided along raised its head four or five feet in the air and its tongue was as long as a man's arm. It made a noise like a train of cars. Its body was the size of a beer-keg, and the track made in the dust was over a foot wide. Great spots larger than a man's hand covered the snake. Colonel Thompson, another Warren county man, says he heard of the monster twenty-five years ago. A circular hunt will be organized as soon as enough men brave enough for the expedition can be got together to rid the community of the monster.

> Albert Lea, Minnesota, *Freeborn County Standard*
> December 2, 1885

Ate Oil Cans

From the Clay City (Ill.) *Advocate*.

Messrs. Tilley, Tooley and Manker, three of Clay City's most reputable citizens, who are engaged in lumbering on Coon Creek, in the western part of this county, had quite a thrilling experience the first of last week with a monster reptile, the genus and species of which are not down in the nomenclature of natural history. On Tuesday afternoon, while squirrel hunting, the piteous howls of Tootsey, the favorite canine, reached their ears. Upon rushing to the spot from which proceeded the howling, they discovered poor Tootsey in the coils of a huge snake that was rapidly proceeding to swallow his dogship. In his excitement Tooley rushed in to rescue his pet, when his snakeship released the dog and attacked Tooley, around whose lower extremities it began to coil itself. A simultaneous discharge from the firearms of Tilley and Manker blew off the reptile's head and he was no more. The snake was of an unknown species, and upon measurement proved to be 9 feet and 2 inches in length. It was of a greenish color with red spots. The stuffed skin of the snake is now in possession of Mr. Manker. The snake was cut open by Mr. Tilley, who found therein three young goslins, two ducks, a young pig, and an oil can which had been missed from the saw mill a few days before.

> San Antonio, Texas, *Daily Light*
> October 20, 1891

King of Reptiles
Julius Hamer Meets a Mad Snake of Gigantic Proportions.

Julius Hamer is a young man of Plymouth, Ill., who despises rats and mice and especially minks, as the latter have been killing his young chickens very rapidly. Accordingly Julius secured a trap and set it in his barn. He thought to catch the mink or other animal that had been robbing his coops, so baited the trap with a small mouse. Last night Julius was awakened by the snorts and stamping of his horse in the barn, where he had set the trap. He rushed out to be met by the horse, who had broken from his stall halter. The animal was whinnying and quaking with fear. Julius heard a great noise in the barn, and he was made aware of a sickening odor rising from the atmosphere. Rushing into the barn his eyes met a sight, by moonlight, which he hopes he will never see again. A monster snake, fifty feet in length, he declares, had been caught in the trap through the neck and the horrid reptile was thrashing around in the barn in a frightful manner. The air was filled with hisses, while the peculiar fetid odor arising from a huge boa when angered permeated all the air. The snake became so desperate that it lashed the stalls into kindling wood and broke out several boards in the side of the barn. Julius gave a terrible yell, which aroused some of the neighbors, but the snake had torn the trap loose from its moorings by this time and started for the creek, which is several miles distant. The men were too frightened to follow the animal far, but they saw the huge reptile cutting a wide swath through the corn as it hurried away. Crooked creek had been very high the last month and it is believed that this snake is the Thompson's Lake sea serpent seen by Rev. A. K Yullis near Lewistown, Ill. The creature has evidently a roving disposition, for it has also been seen in the Mississippi river near Quincy. It is said to be as large as any boa constrictor, but is of a species that inhabits both land and water. Julius is minus his mink trap.

The great rise in Crooked creek also brought untold numbers of buffalo fish from the Illinois river. Squire Jones was out in his feed lot one evening recently when he heard a great flopping down in a meadow piece that had lately been overflowed. He went down to investigate, and found that the water having receded from the meadow had left the ground literally covered with buffalo fish. The entire neighborhood has feasted royally on fish for some days.

Monroe, Wisconsin, *Evening Times*
August 11, 1894

The Boss Serpent.
A Big Snake Located on a Farm in Illinois.

John Phegley of Modoc, Ills., has the boss snake of this country on his farm. This is no snake story, and every statement relative to the Phegley snake can be proved. It is a monster reptile, as large as the heaviest part of a man's leg and 9 feet in length. This is known to be a fact because his snakeship comes out every spring and sheds his skin. Of course the empty shell, not being considered dangerous, is approached and measured, without fear.

The big snake lives in a deep hole or den in the bottom of a "sinkhole" on Mr. Phegley's farm, and while its presence has been proved many times the actual snake has not yet been seen, for he is wary and times his visits in search of food so that he has always kept out of sight, but he is there and has been there for many years.

There are others on the Phegley farm, regular old rattlers, many having already been killed there. Among the recent captures was that of a huge rattler who had caught a squirrel and a tartar at the same time, for the snake was found dead with the squirrel protruding from its mouth. Both had evidently died, the one in eating and the other in trying to keep from being eaten.— Chicago *Times-Herald*.

Lincoln, Nebraska, *Evening News*
July 21, 1896

In April, 2007, a woman photographed a large yellow snake near the shore of Lake Springfield. The *State Journal-Register* (Springfield, Illinois) reported that several herpetological experts were unable to determine the identity of the snake, due to lack of details, though they ruled out the northern water snake. Suggestions included a yellowbelly water snake, a diamondback water snake, or an exotic release. (An albino Burmese python would certainly be large and yellow.) Despite asking people to keep an eye out for it, no one else reported seeing the snake.

Indiana

```
17    3   10
    1     5
    9  2 4
 8       7
   12
    11
     16
     13 6
   14
15
```

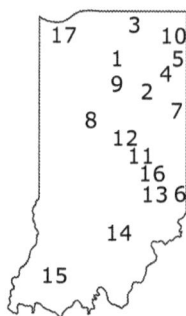

1. Fulton County
2. Grant County
3. Elkhart County
4. Wells County
5. Aboite & Fort Wayne, Allen County
6. Brookville, Franklin County
7. Jay County
8. Frankfort, Clinton County
9. Denver, Miami County
10. Waterloo, DeKalb County
11. Greenfield, Hancock County
12. Hamilton County
13. Decatur County
14. Jackson County
15. Petersburg, Pike County
16. Shelby County
17. Kouts, Porter County

From Fulton Co., Indiana:

V. P. Calvin discovered a snake on his farm which he thinks is about 5 or 6 inches in thickness, and 10 or 12 feet in length. The color is spotted; and the snake is very wild, giving him a very short view of the monster each time, so much so that he has had no chance of killing it. He thinks he first saw this reptile about five years ago.

Rochester, New York, *Sentinel*
August 7, 1875

Some time ago we gave an account of a monster snake seen by V. P. Calvin in Wayne Township. It is supposed that Wm. Montgomery killed the same snake. Some weeks ago he heard his dog barking, and on going where he was, discovered a monster reptile, which he would not undertake to kill without his gun. The dog watched it until he got his gun, and after shooting it he found it to measure just 12 feet. The color of the snake was dark, the head striped, but the kind we have not learned.

Rochester, New York, *Sentinel*
September 18, 1875

V. P. Calvin informs us that the monster snake killed by Wm. Montgomery some days ago does not suit or answer the description of the one seen at various times by himself.

Rochester, New York, *Sentinel*
September 25, 1875

Monster Snakes in Indiana.
Farmer's Fight with the Huge Reptiles

Mr. Joshua Mills, a farmer, lives in the extreme northwestern part of this or southeastern part of Grant county, near the junction of Delaware, Grant and Madison counties. One day recently he was clearing up a patch of deadening on his farm. The day was balmy and pleasant. At ten o'clock in the forenoon he had ready for burning three or four heaps of brush, and when about ready for doing so he conceived the idea of cremating an old log hog pen which stood near by. A hundred feet distant was a well which was uncovered, was operated by a deep bucket attached to a pole, and was used for watering stock. A hundred feet or so further on was a large shallow pond of stagnant water, surrounded by willows and dead flags and coarse grass. The surrounding country is low, wet and swampy. The hog-pen was a fair specimen of ye olden-time stye—roomy, with log flooring laid on the ground and the corners of the structure resting on large logs imbedded in the earth. After determining upon burning the hog pen, on account of the many rats, lizards, etc., he had seen about it, and because of its having done duty so long as to become a pest, he at once proceeded to carry brush and throw into it. He worked an hour at this. When he had almost completed his work, and was in the act of picking up the scattered limbs and throwing them into the pen, he was startled at seeing a monster water snake crawl out from under the logs and start off toward the pond. Mr. Mills told the *Commercial* correspondent's informant that this snake was not less than fourteen or sixteen feet long and about eight inches in diameter. He was so completely dumb-founded that he could not move for a moment, but finally ran toward the serpent, and with clubs succeeded in turning its course away from the pond. He then ran to the house for his gun, an old-fashioned musket.

When Mr. Mills returned the snake was not to be found. He then fired the truck in the pen, and started around firing the other heaps, laying his gun on the ground not far from the hog pen. The burning of the brush on the stye attracted to the scene a horse and colt, the latter a yearling, which were running loose in the clearing. After a lapse of probably twenty minutes after placing the fire to the hog pen, Mr. Mills heard a disturbance among the horses, which were standing near the fire. He looked up. The animals were a hundred and fifty yards distant. The first thing that attracted Mr. Mill's attention was the old horse kicking upward with his hind feet as if wild. The colt was stamping with his front feet. Finally the old horse kicked high into the air and ran, and at the same time something like a long rope whirled into the air and fell to the ground. Mr. Mills saw there was something wrong going on and ran with all possible speed to where the trouble was. The horse, which had run away, stood off a hundred yards distant "whickering." The colt pawed the ground terribly, and it was not until Mr. Mills was right at the place where stood the colt that he comprehended the situation. Around the body of the colt was almost two coils of a large snake. Its tail was four or fives times wrapped around one front leg, its head had moved its coils around the colt's body. The colt's eyes protruded, its breathing was hard, it pawed, pranced and cavorted around as if in the very throes of death. Mr. Mills had no knife with him of sufficient size to cut the reptile in twain. He hadn't time to go and get one, for the colt would die in half a minute's time. The snake was slowly tightening its coils, and on either side of it the colt's flesh protruded, showing the terrible strength of the twist. Mr. Mills took in the situation at a glance. He must burst the reptile with a club, and do it very quick, too. He sprang toward a stake lying on the ground, and just as he was picking it up a whirr! whirr! was heard from behind, and looking up the forked tongue of a murderous rattlesnake presented itself.

The fire was so intensely hot there that it scared the rattlesnake and it sped on and away. But then the colt continued round and round, and pawing, slower and slower, as if ready to fall at any moment. Its very life was being squeezed out of it. Without any delay Mr. Mills rushed that way, and with one fell swoop of his club burst the reptile until its entrails were strewn all over the animal and it fell to the ground dead. Strange, the colt was not seriously injured and it walked away.

But the danger to Mr. Mills had just begun. When the horse ran away, a score or more of snakes of various sizes and kinds were seen running from the fire. A load of shot from his gun failed to kill any of the huge yellow spotted monsters. Owing to the racket caused by the horses the snakes had so far run in another direction, but as soon the animals got out of reach they glided in the direction of Mr. Mills. With the club he held in his hand he succeeded in keeping the snakes away from him for awhile, but finally the number grew so large, and their running so ferocious and promiscuous, that he himself was compelled to retreat. Mr. Mills noticed from a distance that nearly all the reptiles, in their hurry to get away from the scorching flames, were plunging headlong into the well, the others into the pond. From the point where he stood, Mr. Mills saw, he thinks, forty or fifty (...) large snakes of various kinds go into that water.

A comparatively few were rattlesnakes, and they were not very large, but some of the others were huge, fifteen to twenty feet in length. Some of the blue racers carried their heads as high as a man, and swept along like race horses. The sight was one which thrilled the auditer with a sickening horror. When the pen had burned down, so that there was not a probability of there being any more reptiles crawling from it, Mr. Mills approached the well and looked in. The water was probably six feet deep, and the distance to the water from the top about eight feet. From his position he saw some of the most horrible snake fights that can be imagined. Two giants fought until the water was turbulent with mud and blood, spurts of which would occasionally be sent to the top of the well, causing him who was looking downward to start back in a hurry.

Mr. Mills thought to quell the mob by shooting them, but after firing a half dozen loads down into the snakes and quieting a few a new idea possessed him and he at once commenced rolling stones into the well. When nearly filled with stones and wood he cemented the well over with dirt. And now in that grave are buried more hundreds of pounds of snake flesh than can be found in any other hole probably in the State.

Marion, Ohio, *Daily Star*
June 1, 1880

Hunting Monster Snakes with Dynamite in Indiana

Elkhart, Ind., July 27.—For a number of years past an occasional glimpse of a snake of enormous proportions in the vicinity of a marsh three miles south of this city has been the cause of much terror among young and old. Numerous plans have been made to capture him, only to meet with failure.

Yesterday it added to its many previous deprecations by killing a cow belonging to William Hueneryeager, a wagonmaker of this place. A band was at once organised, who tracked him to his lair, where he burrowed in the ground. Numerous efforts were made to entice him by using tempting bait, but he would not come. Finally a dynamite cartridge was thrown into the hole and his snakeship was blown out dead. He measured nearly eighteen feet in length and sixteen inches in circumference. It will be placed on exhibition.

This snake story has the merit of being true, and can be vouched for by a number of our best citizens.

New Philadelphia, Ohio, *Democrat*
August 6, 1885

A Terrible Snake.
He Was So Big that a Man Mistook Him for a Log.

Montpelier (Ind.) *Despatch* in N. Y. Mail.

Great excitement prevails here about a monster snake that was seen by James Baker near the Wells county line. Baker is a business man at Keystone, and his integrity cannot be questioned. He was coming to Montpelier in a buggy and, noticing a black object lying across the road, he naturally took it to be a log, supposing his progress obstructed, but on nearing the supposed log it raised its head apparently about two feet from the ground and made a break for the Fisherback woods, disappearing very quickly. Baker and his two little boys were badly scared, as well as his horse. Baker estimates the snake's length at sixteen feet and as large as a man's body. The snake made a hissing noise that could be heard almost a half mile, and resembled the noise made by escaping steam from a locomotive. Baker described the snake as being coal black except his head, which appeared entirely white, denoting a new species for this country. A gang of men, boys and dogs were quickly formed under the generalship of Uncle Henry Barkman and Bill Howard, the latter having a wide-spread reputation as a snake hunter, and has killed or captured all the large snakes heretofore in this country. The party were armed with double-barrelled shotguns and clubs. They struck his trail late yesterday afternoon, and tracked him for miles through woods and cornfields until night overtook them, and they gave up the pursuit until this morning, when a still larger gang is out with the celebrated snake dog True, belonging to C. J. Maddox. This dog has no desire whatever for other game, and will only hunt snakes, and has tracked and hunted down all the large snakes in this country, and some monster ones have been captured or killed in Blackford county in the last few years. Barkman and Howard took out with them Darby Kritzer, a boy twelve years old, who can charm any snake he runs across.

Syracuse, New York, *Sunday Herald*
August 16, 1885

Aboite Items: The Champion Snake Story

Mr. Henry Vegalcus, while at work the other day in a corn field, was attacked by a monster snake. It chased him about twenty rods, where he seized hold of a mowing scythe, which hung on the fence. With a few earnest attempts he managed to cut the reptile in two, and then killed it. Vegalcus describes it to be about seventeen feet long and about two feet in circumference. It was of a blackish color and its belly was a sort of red.

Fort Wayne, Indiana, *Weekly Gazette*
June 24, 1886

A huge snake "30 feet long and as thick as a barrel" is causing much fright in the vicinity of Brookville, Ind.

Indiana, Pennsylvania, *County Gazette*
October 25, 1893

A Hoosier Snake Story.
The Terrible Monster That Made Jay County Famous.
Portland (Ind.) Letter to The Chicago *Times*.

The torrid weather of the past few days has revived the memory of Jay County's great serpent and refreshed the many wild, weird tales told concerning it.

Probably fifteen years have elapsed since it was first seen in Richland Township. It was on a Summer afternoon and a gang of men were engaged with old-fashioned "cradles" harvesting grain. Near by the field where they were at work ran a little brook, lined on both sides with stubby willows, long, tangled grass, and rank weeds. During the afternoon a boy in the party wandered down toward this spot and was surprised almost into speechlessness by seeing a snake, whose dimensions were to him simply mammoth, dragging its sinuous length along with the rapidity of a race horse. He gave the alarm. A hunting party was organized in short order and armed with rifles, shotguns, revolvers, hoes, spades, and everything available for weapons, the crowd started on the hunt, but were unsuccessful.

The next day 200 men joined in the search, many of them on horseback, with dogs of every kind following. With undaunted persistence they beat to and fro through the bushes, and at several times it was thought that his snakeship had been sighted, but the report proved false. All that could be found was a wide track in the brook's muddy bank and a furrow through the fields of waving grain. The disappointed hunters left for home and for years the snake was supposed to be a myth. Later, however, it bobbed up again and nearly frightened a man out of his wits.

At the time the Lake Erie and Western Railroad Company had a crew of men at work filling in a deep cut at Curtis Hill, eight miles west of this city. The day was a torrid one. A short distance from where the men were at work cattle were quietly feeding under the shade of some scrawny oaks. Suddenly there was a commotion among them, and with heads in the air, eyes dilated, and nostrils distended with fear, they bellowed with fright and started away at a more speedy gait than ever was shown by a stampeded herd on the prairies. One of the men who had been noticing the cattle's queer actions ran up on the bank and was almost paralyzed at the sight which met his eyes.

Twined about one of the oaks was a huge serpent, whose coils enveloped the trunk of the tree in a fold which to man or beast would have proved a fearful death. Its sides were dotted, its vicious little eyes seemed to emit flames of fire, and from the wide-open mouth, with its rows of jagged teeth, darted a long, red tongue.

The spectator was for the time being motionless with fright, but when the immense snake slowly unwound its tortuous length and cast longing glances toward him the power of locomotion returned, and he ran as though pursued by a legion of demons. Summoning help, he returned to the spot, but the snake had vanished, and all search failed to reveal it. To his comrades he described it as fully forty feet long, with a body of prodigious size.

For days afterward hogs and other small domestic animals were missing in the neighborhood, and mothers frightened their children into a restless sleep by tales of the horrible monster which would come and devour them in case they did not behave themselves.

The next place the snake put in its appearance was out in Jackson Township. One dark Summer night it fairly terrorized a gypsy camp. The leader of the gypsies told a graphic tale regarding the event. He said that on the night in question his family and a number of others on their way from Iowa to the gypsy queen's home in Ohio, encamped near a little creek. They had pitched their tents, partaken of the evening meal, and were fast asleep when the neighing of the horses and the barking of the dogs awakened them. Men, women, and children alike were on their feet in an instant, but through the darkness nothing could be seen.

The dogs still kept up their turmoil near the camp, but all at once there was a hush. The frightened curs, with tails between their legs, slunk to their masters' feet in a very agony of fright. An instant later there was a rushing sound, the horses tugged at and broke their tethers, then sped away, with screams of terror that only a horse can give. Huddled together the campers waited, but it was not for long. Through the dim circle of light cast by the smoldering camp fire an immense serpent glided swiftly, with its head fully ten feet in the air, mouth wide open, and wicked eyes snapping death and destruction.

Straight it made for one of the children, and it was almost in the cavernous jaws when the little one's mother, her fears overcome by maternal love, seized a brand from the camp fire and plunged it squarely into the serpent's mouth. The surprised snake whirled quickly, dashing everything aside in its wild course, and glided away in the darkness.

To the gypsies until daylight was an endless watch. It seemed as though morning would never come, and the first gray streaks of dawn were hailed with the joy that a prisoner hails deliverance. All the next day was spent in hunting up their scattered horses. Those who were in the camp that eventful night had a thrilling experience to relate to bohemian friends of the battle with the great Hoosier snake.

New York, New York, *Times*
August 5, 1894

A monster snake, ranging in length from five foot to twenty-five foot, according to the fright of the viewers, has been seen in the neighborhood of Frankfort, Ind.

La Porte City, Iowa, *Progress Review*
May 16, 1896

Reached Out of Sight
This Snake May Be the Sea Serpent on a Land Voyage.

The town of Denver, a few miles north of Peru, Ind., has contained a greatly agitated set of citizens during the past few days owing to the report of a monster snake roaming about in the woods near town, which is made by Eric Gustin, a well known and reputable citizen.

Mr. Gustin tells that he and his wife, together with another woman, were coming home one evening in the buggy and just about dusk they passed through the woods a little to the east of town, when their horse became suddenly frightened and almost fell to the ground through fright. They looked past the animal into the shadows of the trees ahead, and the sight they beheld almost frightened them to death, and the women were made hysterical. It was a monster snake, and its description is thus given in Mr. Gustin's own terms:

"Such a snake I never saw. It lay on a 12 foot log, extending its body the entire length of the log, with its head elevated about two feet, lapping its forked tongue out at us, while mischief lurked in its wicked looking eyes. Its tail reached from the log six feet or more to the fence, and from that through the rails as far back as I could see into the field at the side of the woods. We were in mortal fear of our lives, and how we ever got our horse away from the place in safety I cannot tell."—Chicago *Times-Herald*.

Newark, Ohio, *Daily Advocate*
June 15, 1896

A New Snake
Terrible Monster is Cornered in a Swamp and May be Captured.

Excitement is running high over the discovery of a large snake of an unknown species on a farm near Waterloo. It attacked Allen Lutz, a farmer who was driving a mower, and he had a narrow escape. Lutz describes the snake as fourteen feet long, but others who saw it claim it is much longer.

For two days thirty hunters from this city with guns and traps have been searching the swamps and fields for the snake, and yesterday it was located in a swamp and its capture is expected soon. About two years ago a mammoth snake was seen in this vicinity and it believed this is the same one.

Fort Wayne, Indiana, *Sentinel*
July 22, 1896

Town Hunts for Snake

Greenfield, Ind.—A hunting party, headed by Frank Minor, was organized in this city to go after the monster snake that has been seen in the Brandywine creek bottoms, near this city. The hunt was unsuccessful.

The reptile, which is described as 15 feet long, and with a body as large as a gas main, has stricken terror to the cattle on the farm, and a dog was so frightened a few days ago by the sight of it that the animal ran a mile.

The men who have seen the snake are reliable, and there is no doubt but a reptile of unusual size is at large. It is possible that the snake escaped a circus.

Stevens Point, Wisconsin, *Gazette*
September 2, 1908

Derelict From Circus?
Hamilton County Farmers Tell of Huge Snake That Steals Pigs.

Noblesville, Ind., June 13.—A story comes from the northern

part of Hamilton county which puts to flight anything recorded so far in the reptile line The story comes from a source considered eminently reliable and as all of the county is supposed to be "dry" there is no occasion to doubt it. The Eshelman brothers own a large farm ten miles northeast of this city and they are well known and honored residents of that section. It is claimed that the snake in question makes that farm his home and has been seen there at various times for several years. The last time the monster was seen it was estimated to be five inches in diameter and about fifteen feet long. It is said to be of the Bull species and has been seen to jump a ten rail fence with the ease of a bird. Oliver Brunson, a well-known farmer of the same neighborhood, is said to have lost seven one-month-old pigs, one disappearing each night and the supposition is that his snakeship had a dinner each evening and that pig was his chief diet. Farmers who have seen the reptile are reported to have shot at it, but so far none of them have been able to kill it. The snake was seen three years ago on the same farm.

Fort Wayne, Indiana, *Sentinel*
June 13, 1911

Report This Monster Serpent
Ditches "Flivvers" Along Road

Greensburg, Ind., Oct.—Prominent citizens in the southern (...) of Decatur county declare that tales regarding the depredations of a monster snake are not fables. They declare the reptile, reported to be from 25 to 35 feet long and proportionally large as to girth, has been seen several times in the community during the last few days.

Work has been virtually abandoned in some sections and men, armed with guns, are searching for the monster. A calf belonging to Robert Bishop is said to have been devoured by the snake and the entire community professes intense alarm.

Some residents advance the theory that the reptile escaped from some passing show. Others point to stories emanating from Sullivan and Green counties, many miles distant, of a serpent of similar proportions which is declared to have ditched "flivvers" in encounters on the highways. The Sullivan-Green county reptile was reported at widely separated points within a space of a few hours and it received much publicity—some of it more than half serious—in the newspapers,

Now the Decaturites are wondering if this monster donned figurative seven league boots and emigrated to this county.

Note—It will be observed that in the foregoing dispatch no mention is made of "white mule."

Des Moines, Iowa, *Capital*,
October 8, 1920

Giant Reptile Eludes Indiana Snake Hunt

Brownstown, Ind.—The big snake hunt of Jackson county near here Sunday failed to disclose the monster reptile that has been terrorizing that vicinity. Thirty-five hunters found no snake but did find big imprints showing where a large reptile had wiggled. It has been described as eighteen to twenty-five feet long.

Madison, Wisconsin, *Capital Times*
August 5, 1928

Indiana Party Discover Snake

Petersburg, Ind., Aug. 25. (INS).—A giant snake that has been infesting the Flat Creek bottoms near Cato, in Pike county, near here, for the last twenty years, has been seen again after a lapse of years

Employees of a construction company, inspecting the drainage ditch near the working of the Patoka coal mine, were thrown into a near panic by the sudden appearance of the huge reptile.

The men were so frightened that all had different stories to tell of the monster. Some said the snake was almost a foot in diameter and thirty feet long. Others insisted that it was at least twenty feet long and from nine to ten inches in diameter.

All agreed that the monster reptile had diamond spots on its back and that the scales flashed in the sunlight.

This immense snake has been reported, seen at intervals in that vicinity for the last twenty years. Apparently it travels only at night. Where it crosses the road it leaves a trail similar to one made by the dragging of a small saw log through the dust. Recently the territory infested by this giant snake was purchased by a coal company and sooner or later, it was hoped here, one of the big stripping shovels will unearth its den.

Appearance of the snake recalled the stories of the monster snake reported seen several limes in 1927 in the woods near Sauers which were swept by a forest fire this summer.

Apparently it's hard to keep a big snake story down in Southern Indiana.

Van Wert, Ohio, *Daily Bulletin*
August 25, 1930

Petersburg Fears Snake

Petersburg, Ind., Aug. 10—(U P.)—Citizens here are being extremely careful where they step today. Two mussell diggers, Millard Mead and Lee Cockerham, have reported seeing a snake as long as an automobile which leaves a "footprint" similar to a tire's tread.

The men reported they saw the snake Sunday on the east fork of White river near Rogers, northeast of here. They were paddling down the river when they noticed what they said looked the head of a dog swimming across the river. They kept on paddling and saw the "dog" wiggle up on the bank and stretch out to "15 or 20 feet."

Old timers recall that a giant snake was seen near Algiers several years ago, and that such a snake escaped from a circus at French Lick during the 1913 flood.

Hammond, Indiana, *Times*
August 10, 1937

Monster Rumors Grip Folk of Hoosierland

Indianapolis,—Tales of huge monsters, long snakes, and unshapely reptiles roaming fertile Hoosier soil continued to grow today.

Herpetologists and entomologists notwithstanding, reports of variously described creatures continued to pour in from all parts of the state. Usually, the monsters were reported "glimpsed" by farmers, picnickers, fishermen, or travellers.

Sometimes only the "enormous tracks" of the apparitions were found.

This year's crop of the traditional "monster stories" was touched off at tiny Norristown, a Shelby county community in south central Indiana.

These two farmers were frightened by a "monster of terrible proportions" more than a month ago. The size of the snake or monster grew as the rumors spread thick and fast.

Experts Investigate

Soon, snake hunters, entomologists, and herpetologists (snake experts) converged on the flatrock river in the vicinity of Norristown. Tourists by the hundreds flocked to the area on weekends.

"I'm more afraid of the sightseers than the fabled snake," said Alfred New, a herpetologist who was probing the reports.

The Norristown monster soon became a mobile unit. Reports that he had been seen eminated from points miles from his original location. Residents of Petersburg in southern Indiana thought they had him treed. But he escaped to turn up in three or four other places.

Not to be outdone, citizens of Lebanon, north of Indianapolis, brought to light an "unknown quantity" that rivaled the famed flat rock river monster last week.

This creature made weird sounds like a baby, killed livestock and apparently lived in a gravel pit.

Observers agreed the Lebanon and Norristown monsters were not the same.

Doubt It's Python

Local authorities in Shelby county became interested when it was rumored the flat river creature might be a python. State entomologist Frank Wallace discounted this, said the python was a tropical reptile and the possibility that he would roam to Indiana was "extremely remote."

Meanwhile, the monster stories spread. Sunday night window shoppers in Indianapolis spotted a "huge snake" in a sidewalk grating. Police poked shotguns at the offending reptile, thinking they had central Indiana's monster trapped. The snake didn't budge. It was a novelty ash tray with a stuffed snake on it.

But rumors still flew thick and fast, despite official confirmation.

"I used to tell snake stories myself," said entomologist Wallace. "But every now and then something turns up that knocks all our theories into cocked hats."

Valparaiso, Indiana, *Vidette-Messenger*
August 14, 1946

State Expert is Worried, But Not About 'Monsters'
By William A. Drake (United Press Staff Correspondent)

Indianapolis—State Entomologist Frank Wallace today admitted there might be cause for concern about the frequent reports of gigantic snakes, horrible monsters, and peculiar creatures which have been "seen" throughout Indiana.

The alarming part of the situation isn't the size of the rumored apparitions, Wallace said, but the veracity of some Hoosiers.

Wallace mentioned no names nor particular instances. But, he said, it seemed funny to him that there was never any trace of these prehistoric monsters which have reportedly swept Hoosier farm and woodlands

"Where there are monsters, there must be tracks," he said.

Wallace admitted his interest was mostly with little bugs. As an entomologist he didn't qualify as a monster expert.

Not Much Danger
"But I don't think there's any chance of any Hoosier being choked to death by a boa constrictor," he said.

"These unshapely monstrosities are never reported until two or three days after they are first seen," he said. "Later, they seem to vanish into thin air."

Nevertheless, persistent reports of "enormous tracks" continued to filter in from various Hoosier localities. Farmers, fishermen and tourists glimpsed various sized and shaped monsters in gravel pits, woods, and even on the downtown streets of the state's largest city.

Norristown's "big snake" was still the prize monster. But Lebanon had an "unknown quantity" creature which sounded like a baby and killed livestock. Other areas reported variations.

"It happens every two or three years," Wallace sighed. "Somebody reports a big bobcat or a big snake and then they start from everywhere."

Wallace said the state conservation department had received no reports or complaints from citizens.

"There's always the chance of a big snake being found in Indiana," he said. "The native Hoosier black snake sometimes gets to be six feet. But that's just a starter for most rumors."

Wallace was inclined to agree with a herpetologist who probed the Norristown snake story.

"Water and fear accentuate the size of anything," said the snake expert.

Wallace said he hated to spoil the story but Indiana probably would have to be satisfied with coyotes, a few wolves, badgers and rabbits and squirrels.

Valparaiso, Indiana, *Vidette-Messenger*
August 16, 1946

Hoosier Posse Hunts Big Snake

Fort Wayne, Ind. (UP)—About 100 men, some on horseback, renewed their search Monday for a "sickening blue" snake heaving a head the size of a bulldog.

The snake hunt was called off Saturday by darkness and Sheriff Harold Zeiss said he thought the reptile was "too stirred up."
• • •

He said experts had advised him that big snakes tend to get excited and are sensitive to earth tremors and noise and usually seek a hiding place, until the ground settles down.

The creature was spied wiggling across the road in Fort Wayne State Park by A. D. Crance, his wife and two children. Crance said the snake was "a sickening blue, the biggest one I ever saw."

Hammond, Indiana, *Times*
June 16, 1952

Posse Formed to Find Snake '18 Feet Long'
Indiana Woman Reports Seeing Huge
Reptile Crossing Street in Park

Fort Wayne, Ind., June 14 (UP)—A posse hunted through thick underbrush along Spy Run Creek today for a huge snake "18 feet long with, a head as big as a bulldog."

Sheriff Harold Zeis said he and his deputies found a track left by the reptile as it slithered over the ground in a city park.

But the sheriff said all he caught during the safari was a possible case of poison Ivy.

He has no doubt, however, that the snake is on the loose in this area as reported Friday night by Mr. and Mrs. D. A. Crance and their two children.

Mrs. Crance was driving the family through the park on Indiana Highway 324 when the snake started across the road in front of them.

"It was the biggest snake I ever saw," she said. "The thing raised its head three feet off the ground. It was colored a sickening blue."

Mrs. Crance slammed on the brakes and all four members of the family watched the snake cross the highway and disappear into a grassy field.

The family notified Zeis' office. A preliminary search Friday night disclosed the snake's track, and Zeis organized a fullscale hunt today.

Crance, in describing the snake, pointed to a 14-foot two-by-four timber and told Zeis "it was longer and thicker than that."

Mrs. Crance said she would have run over the snake but she had read some place where a snake was cut in two by a car and the head end flipped through a window and bit the driver.

Robert Snedigar, curator of reptiles at the Brookfield Zoo in Chicago, said the Crances may actually have seen a pilot black snake which is blue-black but grows only six-feet long, and is native to Indiana.

Most people, he said, "always see snakes as longer and more frightful than they actually are."

Anniston, Alabama, *Star*
June 15, 1952

Youth Admits Snake Yarn Hoax
Indiana Boy Faces Hearing in Court

Fort Wayne, Ind., June 19.—(U.P.)

A teen-age boy, whose report that a giant snake "jumped at me" touched off another snake safari here, Thursday faced a court hearing for "disturbing the peace."

The youth said he spotted the big reptile, which previous reports have said was the thickness of a "firehose" and with a head the size of a "bulldog's," slithering through the underbrush near the edge of Fort Wayne.

Sheriff Harold Zeis and his 40-man posse immediately closed in on the area.

After a period of searching in vain, the youth was questioned further and admitted he made up the story.

"I didn't even, see a little old garter snake," he said sheepishly.

The youth was jailed to await his court appearance.

His report was the tenth snake sighting since a family last Friday saw a "sickly bluish" snake making its way across a road in a city park.

Subsequently, another resident and his two sisters spotted a "beige" snake, and two steel construction warehouse workers said they saw the last "six or eight feet" of a large snake slipping into the grass hear the warehouse.

Billings, Montana, *Gazette*
June 20, 1952

Leaders of 100-Man Posse Claim
18-Foot-Long Snake Held in Trap

Fort Wayne, Ind., June 14.—(UP)—Leaders of a 100-man posse said tonight they had trapped a huge snake "18 feet long with a head as big as a bulldog."

The posse, including horsemen and National Guardsmen, closed in on the "blue-colored" monster on the Parker Whiting farm just north of Fort Wayne.

They abandoned searching operations due to darkness, but posted "picket lines" about the area to keep it from slithering free.

Sheriff Harold Zeiss said 100 men would converge on the reptile tomorrow.

Officers had been searching for the snake ever since it crawled across a road in a Fort Wayne park last night before the startled eyes of a couple and their two children.

The posse traced the snake's path through the grass to the Whiting farm. Leaders felt sure they had it trapped tonight because a lot of brush was thrashed around in the area.

Forrest Funk, state conservation officer, said he hoped the men could take the reptile alive. He believed it was either a python or a boa constrictor, a species that prefers to lurk around [...] and marshy boggy areas.

The National Guardsmen members of the 293rd Infantry Regiment of the Indiana National Guard, abandoned maneuvers in the area to aid in the search.

Zeis also called out [...]

Reno, Nevada, *Nevada State Journal*
June 15, 1952

Big-Headed Snake Foils Sheriff Plot on Capture

Ft. Wayne, Ind., June 16—(A.P). A "sickening blue" snake with a head "as big as a bulldog's" eluded hunters for the fourth day in a row today, upsetting Sheriff Harold Zeis' newest stategy.

Zeis believed the 100-man posse, which sought the snake Saturday with planes, horses, trucks and walkie-talkie radio scared it into hiding. Figuring it was holed up in a mile-square area on Parker Whiting's farm, Zeis cut the searching squad to six and ordered them to play a "waiting game."

"Just stand around." Zeis said, "and wait till it pops up some place."

But the snake failed to oblige.

The search started when the D. A. Crance family reported they were out for a drive last Friday evening and stopped to let an 18-foot snake slither across a highway.

The snake was tracked from Franke Park, where the Crances said they saw it, to the farm by slimy trails through fields and brush.

Zeis said a big snake was reported as long as 12 years ago. It may be the same snake, Zeis said.

Long Beach, California, *Independent*,
June 17, 1952

Zoologists May Be Called to Trap Big Snake

Fort Wayne, Ind. (UP)—Sheriff Harold Zeis said Wednesday he planned to call in zoologists to help trap an 18-foot snake with a head "the size of a bulldog's" after two more witnesses reported sighting the reptile.

Zeis laid two steel construction warehouse workers told officers they saw "about six or eight" feet of the snake's tail as it slithered into undergrowth near the warehouse.

The Sheriff said he would question the men further. Meanwhile, he planned to contact zoologists at either the Lincoln Park or Brookfield zoos in Chicago.

Zeis said the last report—the third since the snake was sighted Friday—convinced him the snake is real.

"There's no doubt in my mind now," Zeis declared.

Tuesday Eugene M. Lafavour and two sisters, Violet and Marguerite Schneider, spotted the snake coiled on a road about four miles from where it was first reported. The sisters described it as 'beige" colored and Lefavour said it looked "like a two-by-four."

The sisters refused to leave the car, so Lefavour got out and sent them for help. Lefavour started after the snake but lost sight of it near a stream. When Zeis and his men arrived, Lefavour was standing on the roadside, his face "white as a ghost," the sheriff said.

The snake has defied the efforts of Zeis and his 100-man posse of volunteers, National Guardsmen and mounted deputies to track it.

After Lefavour's report, the men followed the snake's path into the underbrush, but they never caught up with it.

The snake first was seen by the D. A. Crance family on a road in a city park. They said it was 18 feet long with a head like a bulldog's and "sickly blue" in color.

Hammond, Indiana, *Times*
June 18, 1952

Kouts 'Snake' Rivals Loch Ness Monster
By Rollie Bernhart

Kouts — They're seeing "snakes" in the Kouts vicinity these days. This wriggler, described at times as rivaling the size of the mythical monster of Loch Ness, was being estimated at from 15 feet to 25 feet long when I dropped into a Kouts beer dispensary Wednesday.

A quartet arguing at one end of the bar said the monster had been seen by a farmer several days ago while plowing in the vicinity of Burrows Landing near the Kankakee river.

Fortified by more inebriants, the vociferous four insisted that the "snake" fairly covered the top of the tractor when suddenly struck by the vehicle.

He's Snooping

Playing it cagey (I was the snooping reporter in disguise) over a glass of "soft drink," I found out that the unidentified farmer later tracked the wriggling monster northward but lost the tracks in the sand near the Kankakee river.

Taking another sip of my "soft drink," I overheard Corky Maul, of Kouts, verify that he had tracked a snake in the same vicinity about eight years ago. He too had lost the tracks in the sand near the river.

Maul insisted that the "snake" never ventured beyond a mile north of the river. Five years ago it had been tracked again, but to no avail, he said.

Maybe a Turtle?

In the midst of a few more "soft drink" sips, a former deputy game warden told Maul and the quartet he felt the "monster" was probably a large turtle.

This brought vehement protestations from the vociferous four, who decided then and there to do something about it. Armed with a rifle, they jumped into an automobile and headed toward Burrows Landing to hunt down the monster. These men undeniably knew their business. They had seen "snakes" before.

It was also time for me to go. I drained the remnants of my "soft drink." I had gotten my story and felt complacent and satisfied. But to those who see "snakes" I have nothing but sympathy. I prefer to stick to the standard spectre ... pink elephants.

Valparaiso, Indiana, *Vidette-Messenger*
July 13, 1961

Iowa

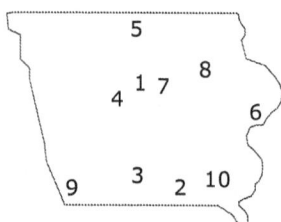

1. Webster City and Randall, Hamilton County
2. Moulton, Appanoose County
3. Lucas County
4. Scranton, Greene County
5. Hancock County
6. Davenport, Scott County
7. Eldora, Hardin County
8. Waterloo, Black Hawk County
9. Imogene, Fremont County
10. Jefferson County

Big Snake Story.
Webster City *Freeman.*

A huge snake, estimated in be from 20 to 30 feet in length and 8 to 10 inches in diameter, was seen a few days since on the Des Moines river, about four miles west of H. Ross' place. There is much excitement in the neighborhood over the appearance of this monster snake, and a large party of men are in the woods to-day endeavoring to find and destroy it. Yesterday it was traced into what is supposed to be its den—a large cavernous opening in a ledge of rocks on the river bank. The story comes to us from a reliable source, and there seems but little doubt of its correctness.

Dubuque, Iowa, *Herald*
August 21, 1880

The Big Snake.

For some weeks past great excitement has prevailed in the vicinity of Moulton, as well as in more distant parts of the state, in regard to the alleged capture of a monster serpent in northern Iowa, by Dr. C. D. Brown. Dr. Brown lives on a farm about three and one-half miles northwest of Moulton, and was said to have the snake in his possession. When the writer visited the neighborhood, the general sentiment seemed to be that the whole affair was a humbug. Men furiously asserted their opinion that there was "no snake there," that if there was one now, Brown had bought one and put it there to conceal the fraud, and that Brown was an impostor generally. All this in spite of the fact that Brown had never taken any pains to assert his claim; had never asked any thing of the public or attempted to attract its attention. His casual statement, corroborating the newspaper statements that he had in his possession a large serpent captured in Webster county, if it was untrue and constituted a fraud, certainly injured no one who attended to his own business, any more than if the statement [was] that he owned a King Charles poodle. But these men did not reason that way—nor any other, very much. Several men were pointed out who said they had, severally, offered Brown $1,000 for his snake, and as a guaranty of good faith had offered to put $500 down. It seemed to the writer quite reasonable that Dr. Brown, not being in the habit of bothering himself much with the affairs about the town, might have some hesitation in rating the financial weight of the parties making the offer at $500, and might have some fear of seeing the offer withdrawn after their curiosity was gratified. It seemed to us, too, that these gentlemen placed too low an estimate on the market value of snakes. We think Dr. Brown is tolerably well posted on the market value of both snakes and men. Others had offered from $10 to $50 for a sight of the snake, after failing to browbeat down a man who cannot be browbeaten. After these men, who had expressed an unfriendly spirit, had failed to budge Brown to favor them with a sight of the snake, they claimed there was none there. Brown's refusal to remove the guards, which he claimed were necessary for the control of the snake, does not seem to argue his insincerity in claiming the existence of the snake. It rather argues a fear of the results of bringing wild men and wild snakes together. One young man, who couldn't have handled one of Brown's legs, was said to have called Brown a liar; another vowed when he started from town that he would shoot Brown if he didn't show up, and conducted himself in Brown's presents in a bulldozing manner. Because these men failed in their efforts at investigation, they decided that there was no snake. All these charges seem to have sprang originally from this class of men. The only "evidence" of the existence of the snake was in the fact that men who had gone in a reckless and insolent manner, and had failed to be favored with a sight of a snake which must be wild, and therefore dangerous and not fit to be exhibited, and is confined in a log crib made especially for trap, and not with a view to exhibition.

On the other hand, we will give the complete chain of facts connected with the capture and subsequent history of the snake, so far as we have gathered them and let each judge for himself. We think the facts are conclusive, and that the reader will agree with us that the reason given by those who raised the cry of fraud were as frivolous as the action of some of them was silly.

Having gathered all the facts we could outside, and being satisfied that Dr. M. Y. Sellers, of Moulton, was the only man in the vicinity who had really seen the snake, we called on him. The doctor seemed to be a little timid about avowing his faith in the snake on account of the ridicule it brought upon him. He was acquainted with Brownhall; treated his family, and was on good term with them all. He had called in Brown's absence, and had expressed a desire to see the snake. Brown's brother-in-law, Houser, who lives in a part of the same house went with him to the cage, which stood in the orchard, and moved some rags from a grated window four or five inches high and a foot or more in length. There was a pane of glass over the grate. The rags were kept in the aperture to exclude the light. The inside of the cage was dark. After looking intently for a while, Dr. Sellers made out the outlines of a short section of the snake's belly, a foot or more, the rest fading away into the darkness. He saw no motion, the snake being kept at the time generally under the influence of chloroform. The doctor was perfectly positive that what he saw was a portion of a live snake, his view of it having enabled him not only to make out the outline, but to tell something of the color and surface of the skin. Dr. Sellers is a man who is careful in forming his opinions, and not hasty in impressing them upon others.

Dr. Sellers accompanied the writer on a visit to Dr. Brown's farm on Friday morning, October 1st. Mr. Pearson, Brown's partner, happened to be on the place. Both were engaged in repairing and fitting up a kind of covered carriage which they had used together in traveling and exhibiting Bible scenery. They received us calmly and courteously, and talked to us as freely as they might have been expected to talk about any other matter. The idea that there was any mystery about the affair seemed to be lost sight of when we came to talk to them. On the previous Wednesday night, they had removed the cage into a cave prepared for it, just back of the house. This cave was about twelve feet square, and was covered over with dirt. The front was composed of two sloping doors, jointed and spouted, to exclude the rain. Dr. Brown unlocked the doors of the cave, and took us in. The cage was set up on trustles, across the back of the cave, and about two feet off the ground. A tube had been put down through the roof of the cave, and was to be let into the cage. Another was to be passed from the end of the cage out through the side of the cave. This was to afford an air passage. This would provide for the escape of the rank effluvia which had made so much difficulty and necessitated so much caution in handling the snake. The cage is the trap in which the snake was caught. It was made for strength, not for show. It is made of split logs, round side out, held together by iron bolts. It is about two feet high, three feet wide and ten feet long. It is lined with inch oak plank, said to be doubled. On one side near the upper right hand corner, as we faced it, is am aperture probably eight inches high and fifteen or eighteen inches long, filled by a grated trap door. This is where the snake entered. Near the middle of that side is an aperture similar in size, but filled with solid grating, covered with a piece of glass as before described. Both these apertures were nailed up at the time of our visit, and we did not insist on having the covering torn off. Dr. Brown said, however, that in about two weeks he thought the snake, being undisturbed, would become so reconciled to his change of quarters and so manageable that a good view of him could be obtained, and he invited both of us to come back about that time. We went into the house and talked with Dr. Brown and his wife and Dr. Pearson. A cabinet marked "Dr. C. D. Brown, homeopathist," explained the origin of Dr. Brown's title. Brown is six feet one or two inches an weights, when in good health, 215 pounds. But when he got home with the snake after its capture, his weight was reduced to 180 pounds, and portions of his cuticle had peeled off. He had been taken violently sick on the way, and the care of the snake had been given over to Pearson and an assistant. These facts are well attested. Brown gives his age at forty-six. He was home in London. Has been in this country about twenty-five years. Ten or twelve years ago, we spent a few months in South Africa. While there, he joined an expedition which went out to capture snakes. He is a naturalist, and has a large collection of specimens. He is not a man of fine education, but he appears to know a little of nearly everything and to be able to do nearly everything. He is a silversmith, a blacksmith, a machinist, a horse-shoer, etc. He is shrewd and not easily read, yet we have no evidence showing that he is not a man of perfect honor. Brown's partner, B. Pearson, of Amey, Iowa, is a man of frank, open countenance; one whom you would trust on first sight. He is a Campbellite preacher, and for several years has made a business of exhibiting Bible scenes. He carries on circulars commendations of the character of his exhibition from Elder A. Hickey, Knoxville, Iowa; Elder G. T. Johnston, Eddyville; M. Chartain, elder of Bethany congregation, Appanoose county,

Iowa; Elder Jacob Wyrigh, Moravia, Iowa; J. Neal, pastor of Christian church, Moulton, Iowa; and Wm. Moore, elder, M. E. church, Moulton, Iowa. Elders Johnson and Leal both testify to Pearson's good character. The writer talked with the Rev. Neal about the matter personally. Pearson is a man incapable of palming off upon the public such a fraud as a bogus snake, which he claims to have seen, to have helped capture, and to now in part own. Brown's wife, also, is a woman apparently incapable of deception, though a woman of fair intelligence. She was present during all of our conversation with Brown when the story of the capture and the subsequent treatment of the snake was narrated. Many of the things spoken of she had knowledge of. Pearson was also present nearly all the time, and took his full share in the conversation. During this conversation they showed us copies of local papers from the scene of the capture, giving the account of the capture, the subsequent arrest of Brown and Pearson, and the later efforts of other parties to capture a supposed mate to the snake. They also showed us letters from parties in the same neighborhood mostly relating to these efforts to capture the other snake. The writer of one these letters happened to be known to both of us.

The Hamilton *Freeman*, in the account we read, spoke of two or three persons who had seen the snake years before. The account which we saw in the Webster county Union was written by a young man named N. L. Turner, who lived in the vicinity where the snake was captured and visited Brown and Pearson nearly every day during the siege. He would be a good man to write to to settle a doubt. Address Dayton, Webster county, Iowa. An old Mr. Bell, a bee hunter in this vicinity was chased by the snake about four years ago, and received such a shock that he never recovered his health, and he died about a year ago.

Last May Brown and Pearson were exhibiting Bible scenes in Webster county. They heard the stories about the depredations of the big snake, destruction of animals, etc., and after satisfying themselves that it was really true, they undertook to capture it. They commenced the search in the curly part of August. When they found it, it was in its den, with its head and three or four feet of its body sticking out. Brown and Pearson were together. The den was in the side of a bluff, over a creek. They stopped up the mouth of the den, built a trustle over the creek, and set a trap (the case before described) on the trustle. Then they built a narrow chute from this to the mouth of the den. The den was probed to a depth of twenty-five feet, where a large chamber seemed to be found. By means of a gas-pipe they injected chemicals into this chamber, to "smoke" the snake out. On the second night after the trap was set the snake entered it. The grated trapdoor, through which he entered, was set with a spring which would be thrown when he crawled backward. It was down when the men went to it. The struggles of the snake had loosened the trestle and broken the connection between the box and the chute. They had to cut a road down the opposite side of the creek to get the cage out. They had plenty of volunteer help. There were numbers of men about there, both at this time and while they were making the preparations for capture. The capture was made on the 19th of August As they camped with the snake in the neighborhood of Dayton, one night, a large mob threatened to hung them, and to burst open the cage, claiming that they were body-snatchers, and had a dead body concealed in the cage. Brown refused to allow the mob to either hang him or break the cage, declaring that he was there to defend his property, just as those men would defend their homes. Then Brown and Pearson were arrested and taken to Dayton. To avoid the trouble of trial and the loss of the control of their property, they proposed that a committee of three men of undoubted integrity and good judgment be appointed to open the case, and make an examination, though they considered this dangerous, in view of the wildness and ferocity of the snake. E. S. Geyer, postmaster at Dayton, Dr. Hamilton and Dr. Gardner were the committee appointed. They put eight ounces of chloroform into the case, took the cover off and looked to their satisfaction, though the sickening odor caused them to abbreviate their inspection as much as possible.

They made oath to the result of their examination: that they saw a live snake not less than thirty feet long, and with a head five to seven inches broad. This is a revealed official proceeding. Any of the committee may be addressed for information. The writer has in his possession a postal card written by postmaster Geyer to J. W. Pulliam, of Moulton, in response to an inquiry. He says the snake he saw was the largest he had ever seen.

Brown and Pearson, accompanied by an assistant came home without further molestation. It may be well to stated here that the snake was fed the night before it was put into the cave. Brown,

Pearson, Mr. Houser, Brown's father-in-law and Houser's father who lives across the way, witnessed the feeding. It was dark, and the snake could not be seen, but the commotion in the cave was evidence enough of its presence. A rooster was put in, and was gobbled up before he could utter a sound. The Housers are respectable people. Two young men named Beggs who were in Houser's portion of the house where there was a party, also heard the commotion. At Albia, on the way home, the snake was seen by a young man named French, who lives at Albia. He was on the fair ground at Moulton, but the writer failed to see him, or to hear his first name.

Dr. Brown thinks this snake is a native of Africa—thinks it probable that he escaped from some menagerie. Here is probably a connecting circumstance, which the writer happened to hear in conversation with George Cramer, of Albia. In 1879, Cramer was managing a circus. Robert Allison, of Wisconsin, owned a cage of large snakes, and in a compartment of the same cage had a crocodile. Allison carried these reptiles with Cramer's circus, and exhibited them in a side-show. Somewhere between Webster City and Fort Dodge, Allison being behind, his cage was overturned and broken in, and all the reptiles made their escape. The crocodile was afterwards retaken, but the snakes were never seen again.

Burlington, Iowa, *Weekly Hawk Eye*
October 14, 1880

A monster snake has been discovered on Charlton river in Lucas county. It is reported as 20 feet long.

Humeston, Iowa, *New Era*
August 8, 1888

A Sea Serpent In Iowa
The Monster that Startled Farmers and Fed on Their Hogs.

A prodigious serpent has been terrifying residents near Scranton, Iowa, this Summer, is stated by a paper of that State. Whether it was a sea monster that had worked up the Mississippi or the Summer resort sea serpent attracted by the World's Fair has not been definitely settled. A Methodist minister, however, is said to vouch for the story, and though since the cool weather set in the great serpent has not been seen, plenty of witnesses remain ready to back up the clergyman.

For some weeks people had been annoyed by what they supposed to be a gang of hog thieves. Jacob Black, a wealthy farmer living ten miles north of this place, had been missing hogs from his pasture, and not until last week was he enabled to locate the thief. One of his neighbors was coming through his hog pasture, which is a large inclosure in the timber, when he was startled by hearing a hog squeal, and, looking up, saw a hog that would weigh about 200 pounds in a coil of the snake and the hog lifted seven or eight feet from the ground.

The man for an instant was paralyzed, as he was only about forty steps from the serpent. The snake seemed to pay no attention to the man, but after smashing the life out of the hog very leisurely wended his way to the river and went over the bank into the same. Allie Griffee, a reliable man and farmer, saw the monster, and when he saw it it was surrounded by a lot of young ones. After the monster crawled away he killed two of the young ones, which are about 8 feet in length and covered with small scales something like the scales of a fish, only coarser.

Several men who were hunting his snakeship found him burrowed in the high bank of the river. There are holes there in the bank which are about 10 inches in diameter, and the snake had crawled into one of these holes, and while his tail was protruding from the hole, which he entered some three or four feet, his head was protruding five or six feet from another hole fifteen feet from the first hole. One of the company, who was about fifty yards away from the snake and across the river, mustered courage sufficient to fire a charge of No. B shot at the reptile. This only seemed to madden him, as he lifted his head high into the air and whistled so shrilly that it sent terror to the hearts of the bravest, and all fled in confusion. They heard an awful splashing in the water and then the whistling ceased. They describe the monster as being at least forty feet in length. They say his head is about the size of a calf's head and the body in the largest places about ten inches in diameter.

One Sunday a large company turned out to see his snakeship. A company of men made an excavation in the bank and found many signs of the monster, but he himself was not there. There were the bones of hogs and young cattle and a portion of one young colt. A reporter, anxious to get a glimpse of the snake, sauntered down the bank of the river 200 yards, when to his right he heard a strange noise, and on looking up he saw, hanging from a huge elm limb, this great, slimy monster. The reporter beat a hasty retreat and was hotly pursued by the snake, and would have been crushed had it not been for Mantie Pattin, who appeared just in time and fired two charges of shot into the snake, which sent him whistling like a steam engine into the river. It is believed that this is a sea monster and that he came up from the Gulf.

New York, New York, *Times*
October 1, 1893

Snakes Big as Logs.
Truthful George Washington Billstein Tells of
Two Monsters Which He Saw.

George Washington Billstein is a man who has never had occasion to take the Keeley cure, the willow-bark remedy, or the snake-bite "resolvent." Those who have known the genial G. W. for these thirty or forty years will take oath that he is in all things temperate, truthful and strictly honest.

George W. had been ever in the Shake Rag country, in Hancock township, Hancock county, partly to see about a horse trade and also to see what the boys were going to do about importing some bloodhounds from Missouri with which to run down a pack of wolves that are annoying the good pioneers of the Crooked Creek country both in Hancock and McDonough counties, says the Chicago *Times*.

There can be little doubt that Van Amberg's circus was wrecked one night several years ago while crossing Crooked Creek en route from Macomb to Carthage. The story has been told that a cage containing two panthers and a litter of young was wrecked by falling from a bridge, and that the animals escaped.

The fact of the matter is that Van Amberg's overland circus came very near being demolished in the storm of that night that caught the caravan as it was crossing the Crooked Creek bottoms. The wind rose and a storm threatened. The performance was hurried to a close, the actors drove on toward Carthage, and the wagons followed. While the wagons containing the cages were crossing the Crooked creek bottoms the storm broke forth in its fury. The creek was high and the cage containing two of the largest boa constrictors ever brought to America got broken and they escaped.

It is not known how many feet long these monsters were, nor how large around, but showmen say the snakes were monsters, and had not yet attained the full size. It is believed that at least one or more of the escaped panthers have remained in this part of the state and are responsible for the awful yells heard by farmers on winter nights and the attendant loss of stock.

But queer stories have been told by the hunters and fishers, who claim to have seen something in the creek at high water that resembled pictures of sea serpents. They hoot at the idea of it being one of the escaped boas, but think that some great snake that inhabits water must have come up from the Illinois river.

George W. Billstein, however, has had an opportunity to test the question, for about 6 o'clock one night, in the Crooked Creek bottoms, he saw a snake that beat Ringling's all hollow. In fact, George W. thought a mighty hollow log had rolled into the roadway, for he did not remember to have seen that log in the road when he drove by that spot that morning:. It was not a hollow log, however, but a monster snake, that lay across the road. His horses gave a snort of fright and reared.

At this the great reptile began to uncoil itself, yards at a time, until G. W. Billstein declares that the woods seemed to be full of snake. The air was filled with a deadening odor that made pioneer Billstein sick at the stomach. No sooner had Billstein got a good focus upon this monster than from a neighboring thicket came another large snake, and then the air was filled with hisses that heat a dozen locomotive explosions.

Billstein's horses were in a panic by this time, and the great reptiles evidently meant business. In a jiffy the half-paralyzed man turned his team and drove madly away, striking another road that brought him home by a longer route.

Telling the story to his neighbors they went to the spot and found great trails in the soft earth, while young trees and bushes had been beaten to the earth as if by a storm. But no traces of the snake could be seen.

Davenport, Iowa, *Tribune*
September 26, 1894

Monster Snake.
With a Body as Big Around as a Water Pail.

The people of this neighborhood are very much alarmed over the fact that a monster snake has been seen at large on the outskirts of the town for a number of days, and it is feared that somebody will be attacked by the serpent before it is killed or captured.

The monster was first seen about three weeks ago by Justice Veltor. Since then it has been seen at intervals by I. W. Valentine, superintendent of the Baptist Union Sunday School, and by Dr. Oliver Jones. The latter chased the snake into the woods in an effort to capture it, but was unsuccessful.

All agreed that the snake is about fourteen feet in length, with a body as big around as a pail. Its head is diamond-shaped, and the top is surmounted by a crest that is shaped like a crown.

Some time ago, Dr. Wood, a resident of this place, died, but before his death he liberated a number of large snakes which he held captive. The snake which has been seen is believed to be one of them. The monster is referred to by the people hereabouts "King of Snakes."

Davenport, Iowa, *Daily Tribune*
September 7, 1895

Snake Chases Man on Horse
A Monster Reptile Near Steamboat Rock Twelve Feet Long.
Other Items of Interest Gathered From the State.

Eldora, Sept. 23.—There has been a monster snake which has been crawling its slimy length through the Van Note pasture, near Steamboat Rock. Lynn Ruby, who has the pasture leased, was the latest person to come in contact with the monster a few days since.

He was riding a pony rounding up some cattle he had in the pasture, when all at once looking over a ravine to the hillside beyond a short distance he saw the snake, which he describes as probable twelve feet in length, from six to eight inches in diameter, with head erect, coming with all possible speed toward him. After taking a. good view of the reptile he concluded he did not care to make battle with him, especially since he had no suitable weapon for successful snake extermination, and turned his pony's head in the opposite direction and putting the spur to him made all possible haste to leave the "critter" behind him in the pasture, although it followed him a considerable distance ere it ceased to give him chase. Ruby says the snake which he thinks had a head broader than the back of a good-sized man's hand, continued to hiss as it followed him.

Waterloo, Iowa, *Daily Courier*
September 23, 1907

Great Snake Lives on an Iowa Farm
Country About Randall Terribly Wrought Up
Two Have Seen Monster
Serpent is Between 25 and 30 Feet in Length
Its Trail Across Newly Plowed Field Measured Eight Inches in Width—Stampede Herd of Cattle.

Webster City, June 11.—Special—Randall, a small town south of this city, and the farm community east is terribly wrought up over the presence of a monstrous snake seen half a mile east of the town. The serpent is between twenty-five and thirty feet in length and its trail across a newly slowed field is larger than the track left by the largest automobile wheel. Measurements of the trail shows it to be about eight inches in width. The presence of this huge snake is well authenticated and people in the community where it has been seen are going armed and are keeping a close watch. It is hoped the monster can be killed within a few days.

The Reptile Described.

H. G. Pederson, a farmer residing east of Randall, has seen the trail of the serpent a good many times, but never knew what it was. The other day, however, while. M. L. Henderson was at his farm plowing in the field with his hired man, Geo. Anfinson, the snake was seen. Henderson was plowing in one field and Anfinson in another, near the H. G. Pederson house. Anfinson saw the snake crossing the newly plowed ground. He could hardly believe his eyes. The serpent's size made an attack upon Anfinson's part [out] of the question. He went over the hillside where Henderson was at work and told him what he had seen. From that distance, which was considerable, the trail of the monster was plainly visible. Henderson sent for a gun and dog. While these were being brought, the men noted a herd of cattle, grazing under a large tree some distance away and in the direction in which the snake had gone, becoming frightened at something and stampede, running in every direction. When the dog arrived he was put upon the snake's trail. He followed it to the tree where the cattle became frightened and thence on down to the Skunk river, where it disappeared. Following the trail back, the men found that the snake had been lying in a clump of bushes, frequented by Mr. Pederson's chickens, this probably being its feeding ground. The hunt of course, had to be abandoned.

No Doubt of Story.

Later, however, in the same afternoon, while back at his plowing, Anfinson again saw the snake. The community is wrought up to a high state of excitement over the matter. M. L. Henderson is one of Randall's most influential and well-to-do business men and Anfinson, too, is widely known as a man of the best standing. The standing of these men and the positions they occupy in the community, is sufficient authentication for the story. The snake is by all odds the largest serpent ever seen hereabouts and a high state of fear is prevalent about Randall, lest someone get into its coils. If not found soon a party will be organized to hunt the monster to its death.

Waterloo, Iowa, *Daily Reporter*
June 11, 1909

Saw a Big Snake.

The report that a huge snake, supposedly a boa constrictor that has escaped from a menagerie, is at large along the Cedar river near Waterloo, is causing considerable apprehension among the timid ones who have a horror for snakes of any description or size. Two young men who were up the river in a canoe Wednesday night reported that they saw the snake as plain as day; that it was a horrid monster, eighteen or twenty feet long; that it appeared to be very hungry and lashed its tail in a menacing way. The young men are active in temperance work.

This startling report finds a counterpart in the recital given by Harry Piersol of Cedar Falls several years ago. Mr. Piersol was fond of life on the river and took his family out frequently in his launch for an evening ride. One night Mr. Piersol went alone for a ride, and was paralyzed by the horrible vision that met his gaze. A few rods away he noticed a commotion in the water, and was astounded to see the head of the huge reptile emerge. It kept coming up and up until about ten feet of the snake was in the air, with the hideous and massive head weaving back and forth, with fire flashing from the eyes, and with an immense tongue, blood red, darting in and out of a mouth as large as an alligator's. Mr. Piersol nearly broke his oars so rapidly did he row away. He does not know how long the reptile was, but it must have been thirty or forty feet, judging from the area of the river that was lashed into a foam.

The snake that has been more recently reported was seen last week at Waverly. At least a similar reptile is said to have been along the Cedar near that city.—Waterloo *Courier*.

Nashua, Iowa, *Reporter*
June 29, 1911

Monster Snake Terrorizing Town
Railroad Men Report That Fourteen Foot Python is Roaming Country Around Imogene.

Railroad men operating on the west division of the Milwaukee report from Council Bluffs that the town of Imogene is in the throes of great excitement over the seeing of a huge python near that town.

For a week or more people have noticed the trail of the monster serpent across the dusty road near the Jim Laughlin bridge north of Imogene. The bigness of it excited the curiosity and wonder of the community. A few days ago, Mike Doyle, jr., saw the snake itself and gives a minute description of the reptile. He says it stretched clear across the road, its head far in the brush on one side and its tail hidden in the weeds on the other side of the road. Its body was about five inches thick and its color was black and yellow.

He puts its length at fourteen feet. He was not at all frightened but the snake was altogether too big for him to attack. Nobody ought to meet that snake unless armed with a Winchester rifle. A company is organizing to capture his snakeship, not that he is doing any harm, although he might swallow a pig or a brood of chicks if he is fond of springs. This is by odds the best snake story in Iowa this year.

Perry, Iowa, *Daily Chief*
September 13, 1913

Snake Story Told By Early Hunter

Snake stories were not infrequent in the early days, and it is believed that the stories and the reptiles themselves grew in length as time has lengthened. The man who told this one vouched for its being absolutely true, but said he did not want his name used, as he did not want to be called upon to prove it.

"One year in the early spring while out hunting, one of the early settlers of Jefferson county came upon the edge of a large swamp. The bank at this place was several feet high. Crawling up so he could see over the bank and not be seen, the hunter waited patiently for some ducks to come in. While waiting there, he was suddenly startled by a commotion in the water. Rising to his feet, holding his gun in his hand in readiness for instant use, a strange sight met his eyes.

"Out in the swamp was a large serpent or snake about twenty feet long, which had a duck in its mouth. The serpent, seeing the hunter on the bank, started to swim rapidly toward him; the hunter, thinking the serpent was making after him, cocked his gun. and when the reptile drew nearer, fired both barrels into its body, without seeming effect. The serpent kept coming toward him at a terrific speed. It looked as if he would surely run head foremost into the steep bank, but imagine the hunter's surprise when the snake seemed to pierce the earth and enter right into the solid bank, disappearing, leaving only a blood stained trail behind.

"After picking up courage, the hunter decided to see where the serpent had gone and found a hole just at the water's edge where it had entered. Whether it died from the effects of the gunshot or not he was never able to find out, but this was the first and only time that he saw this strange reptile."

> Fairfield, Iowa, *Daily Ledger*
> October 2, 1939

Kansas

```
3        1
      4
  5

  6      2
```

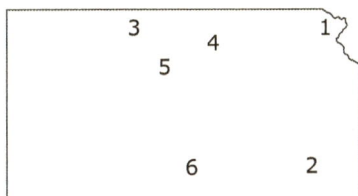

1. Doniphan County
2. Fredonia, Wilson County
3. Phillips County
4. Concordia, Cloud County
5. Luray, Russell County
6. Kingman County

The Luray monster snake is one of the most well-known serpent stories, and continues to be celebrated in that small town (at least occasionally). A 46-foot long replica (loosely created) starred in the 2001 Friendship Day parade. Some of the older residents still remember the stories, with one noting two possible explanations: "that there really was a snake, or there were other towns foolin' with Luray people." (Salina, Kansas, *Journal*, July 7, 2001)

Doniphan County: "According to the testimony of men of veracity the largest snake ever killed in the county, was found by a party of hunters on the prairie some distance southwest of Praire Grove, in the summer of 1861 or 1862. When straightened out alongside a sixteen foot rail, the snake was found to be only a few inches shorter than the rail. It was a bull snake." (Gray 1905)

A Snake Thirty-Eight Feet Long Captured in Kansas

[From the Kansas *Statesman.*]

Readers of the public press will remember a correspondence that appeared in one of the Leavenworth papers in July, 1860, giving a somewhat detailed account of a monster snake that had its abiding place in some of the hollow places of caverns that exist under the eastern mound of Fredonia. The snake had been seen on several occasions; and at one time, as it lay coiled up in the sun, presented a heap about as big around and as high as a whisky barrel. Its skin and spots glistened in the sun, and, on being shot at, rapidly unwound itself, and with head raised from five to seven feet, escaped into its den. It was seen several times subsequently, but could never be captured, although numerous attempts to take the monster had been made. Several sheep and three sucking calves mysteriously disappeared around that period, and the generally conceived opinion was that they had fallen a prey to the monster. The mound, from August, 1869 to about the 1st instant, came to be regarded with fear, and was avoided on particularly hot days as a dangerous place.

Our readers will remember that the first few days of the present month were unusually warm, and one of these days his snakeship came out from his subterranean abode and stretched himself out for a January sunning; but the day proved too cool for his comfort, and, while in a semitorpid condition he was discovered and captured. By actual measurement, it was thirty-seven feet nine inches in length, and around its thickest part measured forty-three inches. It was of a peculiar spotted color, between a blue and a black, with deep, dark yellow spots, and large scales. After it was captured, Drs. Tipton and Barrett had the monster skinned and stuffed, and forwarded it to Wood's Museum, Chicago.

It was too bad that this snake should have been only 37 feet
and 9 inches long. Could he be stretched a little?

Elyria, Ohio, *Independent Democrat*
February 22, 1871

Stock Swallowed by a Snake.
Prohibition Kansas Comes to the Front with a Huge Serpent.

Topeka, Kan., May 13.— The farming community around the
town of Logan, in Phillips county, is in terror over the discovery
of a huge snake that has been committing depredations there for
the last few days. It was first seen last Friday on Crystal Creek,
and is thought to have ascended this small stream from the
Solomon river during the recent freshet. A German farmer made
the discovery. After losing twenty pigs he found the trail of the
reptile and followed it to the creek, where it disappeared. The
next night he lost twenty more pigs and was advised that a neighbor had lost sixty chickens from the same cause. They instituted
a search and claim to have seen the snake and shot at it, but it
managed to elude them by crawling back to the water.

On Sunday it killed a horse, and the farmers reported the matter to the town people and asked for help in subduing the reptile.

A description has been forwarded here in order to determine the
species to which it belongs. It is said to be fifty feet in length, green
in color, with white spots on its body. It raises its head to a height of
ten feet and makes a very wide track where it crawls on the ground.

The facts in relation to the snake are furnished and vouched
for by Bert P. Walker, postmaster at Logan and editor of the
Logan *Republican*.

Decatur, Illinois, *Daily Republican*
May 12, 1897

Sees Odd Snake in Kansas

A monster snake, strange to the state of Kansas, chased Elmer
Gorsuch, truck farmer, from his watermelon field, near Concordia,

A postal cancellation created to celebrate the Luray snake story.

Kans., he asserted. Chased him out not only once but twice and meted out the same treatment to the two Gorsuch boys.

As Mr. Gorsuch entered the field the snake struck at him without warning. He dodged and the snake went after him until he left the field. The reptile is described as being five or six feet long, truly as thick as a large man's arm, head shaped like that of a cobra, and Mr. Gorsuch asserts it has plenty of speed, too. To date, the snake has not been captured or killed. There are no volunteers for the task.

Newport, Rhode Island, *Mercury*
November 10, 1923

Long Snake Story.

Nebraska News Press—Dispatches from Kansas tell of a very interesting and elongated snake, varying in length, with the reports, from 20 to 25 feet. Residents of the community near Luray report losses of chickens and much sleep from the presence of the strange reptile.

Two Clay Center, Neb, men have armed themselves with rifles, knives and a lariat and are venturing into the "big snake country" with hopes of financial returns should they be able to capture the reptile. It is planned, should the expedition be successful, to sell the strange creature to a zoo or a circus.

Lincoln, Nebraska, *Star*
July 16, 1933

Kansas Snake Rivals Monster

(By United Press)

Luray, Kan., Feb. 13.—The Russell County prairies have a monster to rival that of Loch Ness.

Like Scotland's misplaced sea serpent, the Kansas snake has been seen and described by honorable men and women whose testimony cannot be impeached.

James Reiss, a farmer who lives north of Luray, reported the monster more than 20 years ago. He was mowing hay when the sickle bar was lifted 18 inches by a large snake.

The horses ran away, but Reiss stayed around to get a good look at the reptile. The snake traveled across the field "as fast as a horse could lope," was about 25 feet long and had a fan-shaped head with a growth that resembled a cock's comb.

It may be that Reiss' neighbors nudged one another when they heard his story. But about 10 years later Tom Bronson, a Negro, saw the snake. It was crawling from one tree to another, much as tropical snakes crawl from branch to branch of the same tree. The trees were 18 feet apart.

Four years ago, Omar Cochrun, another farmer, saw the snake in a wheat field. That year the wheat grew unusually rank, yet the snake's head was reared above the growing grain. The head was as Reiss had reported 16 years previously, Cochrun said.

One theory is that the snake escaped from some circus that toured Kansas.

Monessen, Pennsylvania, *Daily Independent*
February 13, 1935

Kansas Monster Hoax Confessed By Two Youths

Manhattan (Kan), Oct 29.—(AP)—The myth of the famous "Luray monster" was exploded today.

Leslie Doane of Russell, now a student at Kansas State College, related how he and a companion had manufactured a large "snake" out of canvas, excelsior and paint five years ago and had planted it at various points over Russell and Osborne Counties.

Hunting parties were organized three times in 1931 to track down the "monster," first seen near Luray by a farmer who hurried into Waldo one hot July day with a tale of having seen a snake as big as a stove pipe curled around a wheat shock.

Doane said he and his companion joined in the searches. He added:

We enjoyed those hunting parties more than anyone else, knowing our pet was safely hidden.

Fresno, California, *Bee Republican*
October 29, 1936

In August of 1969, snake hunters were encouraged to chase a large specimen (described by some as up to twenty feet long) in the area near Kingman County State Lake. It was described by one individual as "large and black with faded white spots." A farmer claimed he stopped his tractor for fear of running over it, and a cattleman said his horse tripped over it. (Salina, Kansas, *Journal*, August 3, 1969)

Kentucky

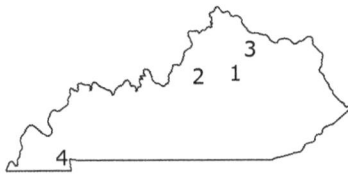

1. Lexington, Fayette County
2. Spencer County
3. Mount Olivet, Robertson County
4. Hazel, Calloway County

Great Snake.—A snake of extraordinary dimensions, was seen near Lexington, Ky., a few weeks since. It is supposed to have been 18 or 20 feet in length, and as thick as an ordinary stove pipe. Many of the neighbors had gone in search of him, and traced him to his cave.—$500 has been offered for his apprehension.

Norwalk, Ohio, *Huron Reflector*
August 24, 1830

—The people of Spencer County, Ky., believe that there is a huge snake, eighteen feet long, in their midst, and there is a tradition that such a reptile escaped from a circus in that county many years since and that it still lives.

Waterloo, Iowa, *Courier*
June 9, 1880

Found Big Snake

Mount Olivet, Ky.—Elisha Highlander, one of Bee Lick's farmers, had an experience lately he will not soon forget. While working in a cornfield on G. E. Linville's farm on Ben Lick he came in contact with a cow snake which, after a spirited combat, he dispatched. It was over 9 feet in length.

Utah newspaper, name unknown
September 13, 1918

Mistook Monster Snake for an Automobile Tire

Paris, Ky.—When Mr. and Mrs. Henry Ross were returning from an automobile trip, they saw what resembled a black and white automobile tire lying in the road. Mr. Ross got out to pick up the "tire" when it moved, and glided away into the bushes. The "tire" was a boa constrictor, the properly of a carnival company that met disaster in a cloudburst near here. Employees of the carnival company later captured the snake.

Olean, New York, *Evening Herald*
February 4, 1921

Snakes Alive, Pistol-Packing Drivers Worse!

Hazel, Ky., (AP) —So many folks are out searching for a big snake in this southwest Kentucky area that Sheriff Woodrow Rickman is getting worried.

He has urged searchers to leave their guns at home. "It would be less dangerous to approach the snake with a switch than to face the wild drivers on the highway and the indiscriminate use of firearms near the scene," the sheriff said.

Hildred Paschall reported he saw a, big snake—possibly a python—while operating a tractor in a corn field June 12. He said it was long enough to stretch across eight corn rows—or about 24 to 30 feet.

Ogden, Utah, *Standard-Examiner*
July 2, 1962

New Chapter Added to Tale of Serpent Monster In Tennessee

Paris, Tenn. (UPI)—A new chapter was added Friday to the growing legend of a shy serpentine "monster" said to be hiding in a vast swampy area near here.

A passing motorist claimed he saw the giant snake—at least 24 feet long according to earlier stories—slithering across the road in front of his car near Hazel, Ky.

A number of others are said to have seen the mammoth snake in past years, including Hildred Paschall, a farmer who lives near the little town of Grassland, Ky., on the Tennessee-Kentucky line. Paschall said he thought it was a telephone pole at first when he spotted it in a corn field about three weeks ago.

The motorist, whose last name was given as "Bivens," stopped at Grassland and told several persons of seeing the snake on Kentucky Highway 893 about 11 a.m. He said he stopped to let it pass.

According to his story, the snake's head and its tail extended off opposite sides of the road at the same time. The pavement was about 20 feet wide.

Rumor has it the snake escaped from a circus years ago and has been making, its home in the creek bottoms about 14 miles from Murray, Ky., north of here. Apparently rather bashful, the reptile only ventures forth on rare occasions.

Paschall said he was applying liquid fertilizer to his corn patch when he saw the snake on June 6, as big around as a stove pipe. He said he halted his tractor and watched the creature for about 20 minutes.

Its head stretched three corn rows to the right of the tractor and its tail was three rows to the left. The tractor itself accounted for two rows. Thus, he estimated, the snake was at least eight rows long, a distance of 24 feet. And it might have been 30 feet allowing for the ups and downs of the field.

It was Brown, he said, with white streaks or splashes and a white spot back of its head,

Mrs. Paschall told curious newsmen her husband is not the sort of man given to tall tales or seeing things that don't exist. Besides, she said, her brother, Hester Charlton, also saw the snake crossing a ford in the creek bottoms five years ago.

She said a man from a snake show came out to investigate and found traces of the snake. She said the snake had never bothered anybody as far as is known.

A search party of about 50 men made a search of the area after Paschall reported seeing the snake, but accomplished little except to trample the farmer's young corn.

A retired biology teacher from Benton, Ky., said from the descriptions, the snake is an Anaconda—a large snake of the Boa family of South America.

Residents of the area, who say there have been reports of a monster snake for the past 15 years, were hopeful that someday it would become as famous as the Loch Ness monster.

Greenville, Mississippi, *Delta Democrat-Times*
July 1, 1962

Louisiana

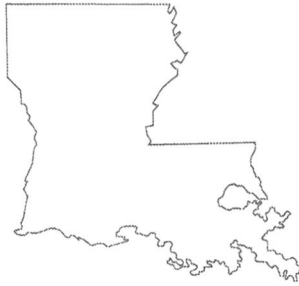

1. Quapaw Bayou, not denoted, location unknown. Mention has been seen elsewhere, so it does not appear to be an imaginary location.

An Immense Reptile
Killed While Preparing to Make a Meal of a Calf
—Thirty-one Feet Long.

Shreveport *Times*.

We were yesterday informed by Mr. Smith, living on Quapaw bayou, that while he and his son William, aged about thirteen years, were out in the woods on Monday afternoon last, driving up their cattle, their attention was attracted by the bleating of a calf some distance from them. Thinking probably that the poor animal had bogged, they started to its assistance. They had gone a short distance down the bayou when they discovered a year-ling, about two years old, in the coils of a huge snake, the body of which was suspended from the limb of a black gum tree, about twenty feet from the ground, and which projected from the bank immediately over the water. Mr. Smith and his son were almost terror stricken at the sight, and stood speechless for several moments, unconsciously watching the movements of the huge reptile as he twined himself around the already dead body of the yearling, and at every coil of the snake they could hear the bones of the calf break.

After coiling itself around the lifeless form of the yearling and crushing every bone in its body, the serpent let loose its hold from the tree, dropped down alongside its victim, and began licking it all over, preparatory, it is supposed, to swallowing it. About this time Mr. Smith recovered his senses, and after watching the monster snake open its capacious mouth several times, he fired on it with his rifle, striking it on the head, and was quickly followed by his son, who discharged a double-barreled gun loaded with buckshot. Both reloaded as quickly as possible and again fired on his snakeship. In the meantime the reptile had coiled itself into a huge mess, and was making a hissing sound that could be heard fully one hundred yards, and was protruding his forked tongue several feet. After discharging about a dozen volleys each, Mr. Smith and his son succeeded in dispatching one of the larg-est snakes ever seen in Louisiana, and probably, North America. It measured thirty-one feet in length, and the body measured ten feet from the head thirty inches in circumference, and about the centre of the body forty-two inches. It has a regular succession of spots, black and yellow alternately, extending from its head to its tail, while either side is a deep purple. Mr. Smith has no idea what kind of a snake it is, but thinks it must be of the boa constrictor species. No doubt this snake has for many years inhabited that section of country and depredated upon the young calves and animals that came within its reach. The skin of this huge snake has been preserved, and will be sent to Shreveport and put upon exhibition.

Fort Wayne, Daily Sentinel
January 12, 1878

Maine

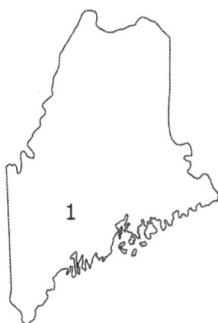

1. Winslow and Gardiner, Kennebec County

The book *All Fishermen Are Liars* (by Linda Greenlaw, 2004), briefly recounts the story of a ship, *Royal Tar*, that sank while transporting circus animals. Besides giving rise to tales of wild animal encounters and strange hybridizations, it supposedly lead to sightings of big snakes on the uninhabited islands near Deer Isle.

The inhabitants of Winslow, Me., on the banks of the Kennebeck River, are very much excited over the appearance in that town of a huge snake. A few days ago, Mrs. Smiley was walking in the woods in the vicinity of her home, when she saw what she at first supposed to be a large spotted snake of the common variety, frequently seen in the fields, but which was really only the tail of a reptile which, she says, must have been 8 or 10 feet in length. The snake was coiled around a tree, and his head, with distended jaws and forked tongue darting in a frightful manner, was projecting over a limb above Mrs. Smiley's head. The lady ran for life. A neighbor's house was but a short distance away, and, rushing in, Mrs. Smiley fell fainting upon the floor. She says the snake followed her nearly to the house. The Sunday following (which was last Sunday week), as a son of Deacon Palmer was walking to church, his attention was attracted to the stone wall beside the road, and on the top of the wall, basking in the sun lay a monster snake and two smaller ones. Mr. Palmer says the largest one must have been 10 feet in length. He ran back a short distance to procure assistance, but when he returned, the reptile had disappeared. Several parties have been out hunting the snake for the past few days, but without success.—Kennebec *Journal.*

Decatur, Illinois, *Daily Review*
November 3, 1878

A certain man in Gardiner, whose word is as good as gold, we are gravely informed, tells of a huge snake which he saw near the Pickering farm on the Brunswick road. In his judgment, the snake measured ten feet in length and was larger round than his arm. Some people might question as to this man's sobriety and have an idea that his sight and power of imagination were augmented by the too free use of stimulants, but such is not the case as he is said to be a temperance advocate.

Bangor, Maine, *Daily Whig and Courier*
August 15, 1895

Maryland

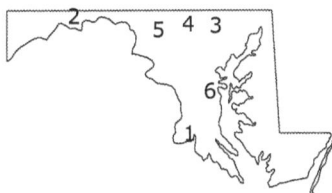

1. Blenheim plantation and Nanjemoy Creek, Charles County
2. Bedford Valley near Cumberland, Allegany County
3. Hall Spring, Baltimore County
4. Bachman's Valley, Carroll County
5. Brunswick and Middletown, Frederick County
6. Dodson, location uncertain, possibly Anne Arundel County

Stephens (1988) noted that Nanjemoy Creek, Charles County, had a history of big snake folklore up through the 1960s.

One mystery animal I won't detail here is Chessie, which as the late Mark Chorvinsky (publisher of *Strange Magazine*) noted, is very much an amalgamation of various strange animal sightings, including cetaceans, sea turtles, strange fish, etc. Certainly, there is the occasional snake-like story. The Frew tape certainly points to a snake-like animal, and there is quite a bit of investigation still to be done there. A Maryland group, the Enigma Project, had worked on that for a while, and I hope that research continues in that direction.

A Snake Story

In an unfrequented part of the town of Blenheim, in this county, is a large range of rocks, extending several miles along the north side of a steep mountain. In these ledges are several openings or caves, the interior of which never has, and since a recent occurrence, probable never will be explored by any human being.

Some time last month, two young men, the eldest about 20, and the other 15, were returning from a fishing excursion, by one of the most considerable openings in the rocks. Being somewhat fatigued, and not apprehensive of danger, the largest boy seated himself several paces from the cave while the younger one stood a rod or two in the rear. Getting into conversation, the oldest boy turned his back to the cave while the younger faced it. Directing his eyes towards the cave, the latter observed the head of a huge monster, issuing from the den. He was so much frightened, that without giving the alarm to his companion, he sprung instantly upon an adjoining rock, and after recovering a little from fright, called out to the other who till then maintained his sitting position, with his back to the cave, wondered what could have occasioned his sudden flight. At the moment he turned his head and for the first time, saw the monster within a few feet of him; his head elevated some 6 or 8 feet from the ground, ready to pounce upon him. He gave two or three dreadful screams, and we may suppose, made good his retreat to his companion. The monster no way daunted, continued his course down the hill some distance, turned back and entered his den. During this time the boys had a fair view of him; and they assert that after making due allowance for fright, they are fully convinced that his length exceeded thirty feet, and his body as large as a common saw log, covered with irregular spots, about the size of a man's hand, of bright red and jet black.

The next day, the oldest boy whose name we believe is Sanford, and who is said to be no ways deficient in courage, went alone to the den, and seated himself on a rock that projected over the entrance, where he had not remained long, when his snakeship made his appearance—descended the declivity further than before, breaking the old sticks and limbs as he passed over them, until Samford made some noise from above, when he turned his course and again entered the den.

The alarm was given in the neighborhood and a number of people collected at the den, but they were not gratified with a view of the monster; his course, however, could be plainly traced; and where he had passed over the sand or leaves, it appeared like the trace of a large log. Since these facts have spread the cave has been visited by great numbers of people from the adjoining towns, and the facts above stated have been communicated to us by several of the most respectable people in the towns of Blenheim and Jefferson, many of whom have visited the cave, and seen the track of the serpent.

They state that they have the utmost confidence in the veracity of the young men who saw him; and as additional confirmation, they say that a very offensive smell, similar to that of large snakes, have been observed by all who have visited the place. *Village Herald.*

Hagerstown, Maryland, *Mail*
July 3, 1829

The Bedford Valley Snake.—Two of our citizens have visited the spot where this huge serpent was seen, with a view of capturing the monster. They were unsuccessful; but received abundant evidence of his actual existence. They saw and examined the skin he had shed, and found it fully twenty-one feet and six inches long. They also saw and conversed with Mr. John Elder, a most reliable citizen, who had met the animal face to face. Mr. E. encountered him in a lane, across which he was lying, with his tail in one meadow, and his head near the second fence. From his dusty brown color, Mr. E. mistook him for the ridgepole of the fence, until his horse started back with fright, when the serpent reared up to the full height of the rider and darted fire from his eyes. The horse instantly whirled and dashed off in alarm, and by the time he could be brought back to the spot, the snake had disappeared in the high grass. Mr. E. thinks he is between 20 and 30 feet long. Barnum may get him yet.—Cumberland (M. D.) *Journal.*

Fond Du Lac, Wisconsin, *Union*
September 29, 1853

A Snake Story of Huge Dimensions—Considerable consternation exists at present among the people residing in the vicinity
of Hall Springs, on the Harford road, about two miles from the
city, owing to the stories extant about the appearance of a very
large snake in the neighborhood. Those who think they have seen
his snakeship say that he is fully fifteen feet long, about eight
inches in diameter and two feet in circumference. On Tuesday
last he was seen, it is said, on the place of Mr. J. F. Lee, the owner
of the Olive Mill. The snake passed through the garden, and came
within a hundred feet of the stable. Its track was afterwards
measured by Mr. Lee, who states its dimensions to be eleven and
a-half inches in width in the narrowest place and about fifteen
inches in the widest. Mr. Lee's place is a short distance beyond
Hall's Springs, and is located on the east bank of Herring Run.
On the west bank opposite the dam there is a large rock, with a
sort of cave underneath. This is thought to be the den of the monster snake, who has been noticed on several occasions in the
woods near by. The miller in the employ of Mr. Lee says that a
few days ago he noticed the snake lying, apparently asleep,
under the breast of the dam. He apprised Mr. Lee, on the latter's
return from the city, of the fact, who scarcely believed the story.
The visit of the snake to his place on Tuesday last, however,
thoroughly convinced Mr. Lee that the snake in reality existed,
and was one of unusual proportions. This gentleman avers that a
snake of the size described has the capacity of swallowing a calf
four or five weeks old, and that a boy or a girl ten or twelve years
of age would be a mere mouthful. Mr. Lee states that he is willing
to give $500 for the capture of the snake alive, or in lieu thereof
$25 to any one who will kill it. Several attempts have bee made to
kill the snake, and have failed for the simple reason that the
parties making the attempts on coming across it and noticing its
huge proportions incontinently fled. The Superintendent of the
Turnpike Company, who is a portly German, was walking in the
woods a few days ago when he met the snake. He rushed from the
spot, and never slackened his speed until he reached the door of
his domicile. His nervous system was so terribly shocked that he
has not yet entirely recovered. Several other persons have seen
the snake, and the stories that are afloat about it have been
productive of a great deal of excitement and comment in the
neighborhood of Hall Springs. It is understood that an effort will

be made in a few days to capture the snake. It is proposed to fasten a strong chain, with a sharp hook attached, to a tree in the neighborhood of the den. On the hook will be placed a live chicken. The snake, it is confidently expected, will go for the chicken, and take in the sharp hook as well, which, it is thought, will insure its capture.

Baltimore, Maryland, *American*
July 2, 1875

The Hall's Springs Anaconda
Its Capture and Escape.

It will be remembered that a few weeks ago the American gave a description of a monster snake, which it was claimed had been seen in the vicinity of Hall's Springs, on the Harford road. Several parties have been organized in the city for the purpose of hunting his snakeship, but most of them on going out to Hall's Springs and hearing the marvellous stories told about the serpent, concluded that discretion was the better part of valor, and that they would let him severely alone. They were told that the snake was fully sixteen feet long, that he could swallow a calf several weeks old, that a four year old child would only make a half a meal for him, that he could travel faster than the fleetest horse in the county, and that when in pursuit of an object he could shoot over the ordinary country fence or stone wall with as much ease as he could make his way through the grass. In fact, the most wonderful stories are told about this snake, and it is safe to say that in a month's time his size will have increased in the minds of the many of

The Excited Denizens
of Harford road to at least thirty feet in length, and that he will be full able by that time to swallow a whole cow, horns and all, without any choking. On Saturday there were rumors in Baltimore that the snake had been caught and had afterwards made his escape. The rumors were tracked to the Baltimore Corn and Flour Exchange, of which Mr. Lee, of Messrs. Lee & Shaw, is a member. Mr. Lee is the owner of the Olive Mill, situated a short

distance beyond Hall's Springs. This gentleman resides near the mill, and a reporter of The American visited him yesterday for the purpose of ascertaining the very latest concerning the snake. Mr. Lee stated that

The Snake Had Been Captured

in a trap on Friday night or early Saturday morning, and had freed himself from the trap on account of his great size and strength by breaking a portion of the top off. He accompanied the reporter to where the trap was set. On Saturday morning after the escape of the snake, the trap was reconstructed. It is now about eighteen feet in length. The sides are made of boards one inch and a quarter in thickness, while the top and bottom are made of strips of the same material and thickness. These strips, which are about six inches in width, are nailed firmly on, and around the whole have been placed pieces of thick hoop iron. The opening is fourteen inches square, being sufficiently large to admit the snake, which is supposed to be about eight inches in diameter and two feet in circumference. In one end of the trap there is a box twelve inches square containing a live chicken In the other end of the trap there is a door sliding up and down in grooves. This door is kept open by means of a lever which works on a pivot, and is attached to a trigger. In passing town the trap to secure the chicken the snake will touch the trigger, when down comes the door, and he is a prisoner. It was in this manner that he was caught on Saturday morning, but finding himself imprisoned, he got his back up, so to speak, and three or four of the slats composing the top flew off and the snake was once more free. Mr. Lee says he visited the trap about 5:30 Saturday morning, when he was at once convinced from its condition that the snake had been in it during the night or morning. Besides the fact of the top of the trap being broken, there were two distinct tracks along the ground, marking the ingress and egress of the snake. At about 8 o'clock the same morning a colored man walking down a path on Mr. Lee's place, encountered the snake, which was lying cosily on the race bank. The colored man did not stay long enough to get an accurate idea of the size of the snake, which will account for his statement that the body of the snake was as thick as a flour barrel and twenty feet in length. The snakes had not been seen before that morning for two weeks. He is supposed to reside on a small island, which is covered with a dense under-growth, and is situated about a stone's throw above the Olive Mill, on the east bank of Herring Run. He has been seen going in that direction. Mr. Linhardt, Mr. Lee's miller, it is alleged, has seen the snake several times. A laborer named Thomas Rickert, who was fishing in the dam several days ago, also encountered his snakeship, who raised his head and stuck out his forked tongue. Rickert flew from the spot and was unable to work for two weeks afterwards.

Five Hundred Dollars Reward.

Mr. Lee considers the snake harmless, unless attacked. He is very desirous of capturing him alive, and would give $500, or even more, to any one who would bring about this result. The new trap is like the old one, with the exception that it is bound with hoop-iron. The talk occasioned by the reports about the wonderful snake has deterred many persons, it is understood, from visiting Hall's Springs woods, a well-known place for hold-ing picnics. There is no danger whatsoever of a visit of the snake to any part of these woods, as he does not go but a short distance from his island home. Passengers on the cars yesterday to Hall's Springs talked of little else save the snake and some of the stories told about him throw the tales of Baron Munchausen completely in the shade.

Baltimore, Maryland, *American*
July 26, 1875

The Monster Snake.—The excitement about the monster snake, which has been seen near the Olive Mill of Mr. Lee, about half a mile from Hall's Springs, on the Harford road, continues. Mr. Lee has increased his reward of $500 for the capture of the snake alive to $1,000. Mr. Lee is a man of means, and evidently means business. Mr. Frederick Weber, the brewer, living on the Harford road, and other persons residing in the neighborhood, who disbelieved in the current stories about the snake at first, have visited the ground, talked with those who have seen the serpent, and are now convinced that the reports have a foundation in fact. The $1,000 reward ought to induce some of the adventurous spirits of this city to make an effort to capture his snakeship, and thus secure a snug sum of money and relieve the apprehensions of the mothers of the vicinity of Hall's Springs, who are fearful that some day the snake may swallow one of their children. It is claimed that his enormous size fully admits of such an event occurring, but if it ever does that snake will be prevented from his increased size from entering the trap set for him by Mr. Lee.

Baltimore, Maryland, *American*
July 27, 1875

The Monster Snake.—The Hall's Springs anaconda still continues to be one of the principal topics of conversation among the residents of that locality. The huge snake has not been seen since last Saturday a week, when it was trapped and escaped. The live chicken which is kept in the trap as bait is fed daily, and is in as good conditions as its narrow quarters will permit. Mr. Lee, the owner of the mill near the small island where the snake is supposed to spend his hours of repose, has employed two men to keep constant watch. These men are paid so much per week, and, if they succeed in capturing his snakeship alive, will receive each $500 additional. No small children have been missed yet by the residents of the neighborhood, though the snake, from its alleged size, could make away with one a week without any trouble. The escape of the children may be accounted for, also, by the fact that some snakes don't feed only once every three months, and that the Hall's Springs fellow has not got hungry yet. Mr. Lee thinks that, from his size, one square meal a year would do him. He also states that, though he has never had the pleasure of seeing the snake, he thinks, from what he can hear from those who have, that he is a black snake. The bite of a black snake, he states, is harmless, but his hug is fearful. Black snakes are only wicked in the spring of the year, so that the Hall's Springs people can take comfort, as they have several months of comparative safety before them. There is a small private path leading from Stevensville through Mr. Lee's property to the Harford road. Several of the colored people were in the habit of using this path until about a week ago, when one of them met the snake. The [man] did not run—he simply flew back to Stevensville, leaving

the snake undisturbed possessor of the path. The colored population of Stevensville avoid the path now. Mr. Lee and a couple of
friends a few days ago went to the island home of the snake for
the purpose of hunting him up. They saw his tracks plainly in the
underbrush and swamp. The tracks were of the same dimensions
as if made by the trunk of a good sized tree. The search was unsuccessful.

Baltimore, Maryland, *American*
August 9, 1875

The Hall's Springs Snake.—The monster snake who lives and
has his being in the vicinity of Hall's Springs, on the Harford road,
still remains uncaptured, but will not retain his liberty long if the
assertions of H. B. Hedges are to be believed. Hedges is from the
West, has caught snakes for menageries, and is willing to catch
the one near Hall's Springs for Mr. Lee's one thousand dollars
reward. Hedges proposes to go out to Mr. Lee's place, camp near
where the snake is supposed to be when he is at home, and wait
for something to turn up. He will carry with him all the appli-
ances of the professional snake-catcher, and avers he will not leave
the spot until the Hall's Springs anaconda is a prisoner. The
serpent has been keeping on his island home of late, though he
took an airing last Sunday. Mr. Lee at 9:30 A.M. saw his track
plainly marked near the door of his stable. The track could be
followed very easily for some distance until it crossed the breast
of the dam.

Baltimore, Maryland, *American*
August 20, 1875

The Huge Snake Swallows Two Pigs.—The monster snake of
Hall's Springs was on the rampage on Thursday, and his exploits
have since been the talk of the neighborhood. Charles Knox, a
gardener, has a place about half a mile from that of Mr. Lee. On
Thursday morning his hired man saw the anaconda in the vicinity of the pig pen. The man was so paralyzed from fear that he
could scarcely move. While gazing at his snakeship the latter

seized a large turkey, and was making preparations to swallow him, when the turkey succeeded in getting away minus a number of feathers and a portion of his tail. The snake then moved off. On looking into the pen it was discovered that two of the four week old pigs had been swallowed by the snake. At least this was the only way in which their sudden disappearance could be accounted for. When this fact became generally known some consternation prevailed, as it was thought that if the snake could so easily get away with two pigs he might dispose of a stray child of the right size. The fact of his having swallowed the pigs may account for the snake's indisposition to make any effort to recapture the turkey. Some anxiety prevails as to what will be the ultimate result of such a monster being allowed to prowl around, and several are of the opinion that the county authorities ought to offer a suitable reward for his capture.

Baltimore, Maryland, *American*
August 21, 1875

A monster black snake, so old that it has become rusty and moss-grown, has made its head-quarters in Bachman's Valley, and has been several times seen the present summer, by different persons. Efforts for its capture are on foot.

Hagerstown, Maryland, *Herald and Torch Light*
September 29, 1875

A Regular Constrictor.

Mr. Chas. Thomas, a 16 year-old lad, while picking berries near Brunswick several days ago, encountered a monster black snake, which he killed with a stick of wood and which measured 13 feet 4 inches in length, and as thick as a man's arm in the centre of the body. It took some nerve to tackle a snake like that.

Frederick, Maryland, *News*
July 26, 1893

Snake 12 Feet Long
Monster Reptile Was as Big Around as a Quart Measure.

Middletown, Md., July 14—Several days ago while Messrs. Roy Gross and Ralph Bere, employed by Mr. Luther Kefauver, on the Lloyd M. Kepler farm, were cutting away underbrush to get to a place to make a fence, they suddenly came across a monster black snake. Gross says he sent Bere to the house for a gun because it was too big to tackle with a club, but before Bere got back with the gun, it had crawled out of sight. They carefully searched or it and found it in the underbrush, then, Gross says he shot it about three feet from the end of its tail, and as its head was past the second panel of fence, Gross says it must have been over 12 feet long. He says it was as thick as a boy's leg, or as big around as a quart measure.

After shooting the reptile they beat a hasty retreat, fearing if it was wounded by the shot, it might give battle. They refused to finish the fence job. Neighbors will take up the search in the hope of killing the big snake. Gross says he is sure this is the same snake that he saw in the hollow of a large poplar tree last spring in the dividing line of the John H. Routzahn and the Kepler farm, at which time he went to the house for his gun, but found it gone when he returned.

Frederick, Maryland, *Evening Post*
July 14, 1911

There is a big excitement at and around Dodson, Md, caused by a monster snake. The women and children have barricaded themselves inside their homes and terror reigns supreme. The men are all out with all the available arms of every description. The monster has been shot at innumerable times but from what can be learned, the shots had no more effect than pouring water on a duck's back. From description given by eye witnesses the snake is about twenty feet long with a yellow head about six inches broad and the body is about eight inches in diameter. The monster is causing a great deal of uneasiness for fear it will destroy some of the children before it can be captured or killed.

Keyser, West Virginia, *Tribune*
July 12, 1912

Massachusetts

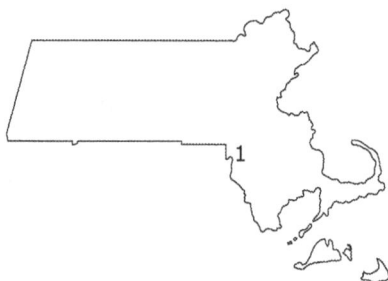

1. North Attleborough, Bristol County

Great Snake Story.

There is a letter in yesterday's Providence *Chronicle*, signed
Joshua Buddington, dated at North Attleborough, and giving a
minute description of a huge serpent, which Mr. Buddington says
he killed in one of his pasture lots, where the reptile had killed a
cow, around the hind legs of which it was coiled. The snake was
supposed to have been in the act of extracting the milk at the
time of their arrival. A severe blow from a heavy club killed him.
The writer adds—

"This, we venture to say, is one of the largest, if not the longest, snakes every killed in this country. Of what species I have

not yet been able to ascertain. His back is Zebra striped and the belly of a dark green, with small blacks spots thickly interspersed.—Around the neck and directly back of the jaw, are four stripes or rings of a bright yellow color, and just under the throat, a small bag or hollow membrane is suspended, filled with a thin liquid substance. This membrane is perfectly transparent and through it the appearance of the contents are dark green. The length of the snake is 14 feet 8 inches—circumference around the largest part of the body, 1 foot 10 1/2 inches—from the end of the upper jaw to the eyes 5 inches—width of the head which is very flat, 7 1/4 inches.—It is now lying in my barn buried in salt. As soon as I can procure a suitable person, I design to have him skinned and stuffed, after which he may be exhibited in your city, and I can assure you he will be well worth viewing."

Wellsboro, Pennsylvania, *Tioga Eagle*
August 2, 1843

Michigan

1. Hastings, Barry County
2. Salem township, Washtenaw County
3. Lansing, Ingham County

Lake Monster.

Kalamazoo, Mich., July 21—The big snake which is said to exist in the neighborhood of Carter's Lake, Barry county, and known as "Carter's snake" has been seen again this week, this time by Henry Marble. The snake is variously reported as fifteen to twenty feet long. Marble was terrified and came to Sexton McElwate's with his horse on the run. Hunts have several times been organized to kill the snake without result.

Cedar Rapids, Iowa, *Evening Gazette*
July 21, 1894

Hunt Snake With Chloroform
Hastings Farmers Will Make Second Attempt to Kill Monster.

For the last fifteen years an Asiatic boa constrictor has been inhabiting the woods, two miles north of Hastings. The reptile is, according to reports of those who have seen it recently, nearly twenty-five feet in length. The huge serpent thrives by devouring sheep, pigs, calves and chickens. Ten years ago a meeting of the residents of Hastings was called, for the purpose of hunting down and killing the reptile. There were many volunteers for the expedition, but after getting one sight of the huge monster nearly all turned back, and the snake was allowed to continue his existence. The serpent was seen a short time ago, after a long absence, and a number of residents have signified their willingness to start out again, not to return until the snake is killed. The hunters will be armed with shotguns, axes, sledge hammers and other weapons of defense and torture. It is proposed to take an enormous supply of chloroform, to be used if necessary.

Bessemer, Michigan, *Herald*
June 29, 1907

Michigan 'Reptile' Object of Search

Salem, Mich., June 17 (UP)—A 30-man posse oiled up guns today for an all-out hunt for the fabulous snake of Salem township.

They plan to begin Sunday on the farm of Mr. and Mrs. Robert Lewis, who measured the tracks of the "monster reptile" and figure it is 14 feet long and "about six inches around."

Mrs. Lewis has kept their two youngsters, Bob, Jr., 10, and Barbara Ann, 6, inside most of the time since the first impressions were spotted about three weeks ago.

A neighbor, 81-year-old George Bowen is the only person known to have seen the snake.

"The thing reared its head out of the tall grass at least three feet the other day and scared my horses half to death," Bowen said. "It was so big around I thought it was a log until it started to chase me."

Holland, Michigan, *Evening Sentinel*
June 17, 1949

Michigan Town Hunts Great Serpent
Seen On Farmlands

Salem, Mich.—(AP)—Salem townsfolk are convinced there's a great serpent still slithering about this farmland community about 30 miles west of Detroit.

A posse of Salem citizens—100 men strong and armed with an assortment of weapons—set out to get the monster yesterday. But all the hunters bagged was two blue racers and eight garter snakes.

Stories of the serpent started circulating nearly a year ago. Then, a local farmer reported, a huge snake—some 17 feet long— appeared while he was driving his tractor, and it outdistanced him when he gave chase.

Three weeks ago 80-year-old Farmer George Bowen gave the serpent-terror a shot in the arm, reporting he saw the animal on his land.

"Its head was standing four feet off the ground," Bowen said. He reported he hurried to his house for a shotgun, but when he returned the serpent was gone.

It was Bowen's land, some 30 acres of marshy, mosquito-ridden ground, that the posse searched. But if the big snake was there, members of the posse said, it certainly has moved by now.

While only two persons report actually seeing the serpent, dozens of farmers can tell about missing chicks, ducks and even a young calf.

Mothers have been keeping a close check on their children, particularly warning the youngsters to stay away from the home of Carpenter Robert Lewis, who located the "serpent's nest" on his property.

Lewis said the nest is four feet wide. Two paths lead away from it, he added.

Some of the local folks have a theory that the snake is one that escaped from a circus in Ann Arbor 30 years ago.

But Conservation officer Davey Crockett of Ypsilanti doubts that. That serpent was a tropical one, he explained, and would have found the going rough in a Michigan winter.

Sheboygan, Wisconsin, *Press*
June 20, 1949

No Snakes in Ireland But Oh Boy, Look What's in Michigan!

Salem, Mich.,—(UP)—The Salem serpent—variously described as a reptile 17 feet long and six inches thick which frightens horses and outruns tractors—was on the loose again today.

A posse of 100 fanners searched the marshes yesterday after a farm woman reportedly found the serpent's nest which it had deserted with the coming of warm weather. All they found in the mosquito infested swampland were two blue racer snakes and eight garter snakes.

The search started from a "nest" four feet wide which Mrs. Robert Lewis found on her husband's farm. The nest had two paths leading from it through the grass, but neither led to the serpent.

The creature was last reported by George Bowen, 78, who said he saw it in the grass two weeks ago when a horse he was unhitching began acting up 'like a swarm of bees had lit on him.'"

"The serpent had his head about three feet off the ground and his fangs were something terrible to see," Bowen said.

He added that the snake's head was five inches thick.

This is the second summer that the serpent has been reported in the Salem township.

Last year a farmer said it out raced him on a tractor. It is blamed for disappearing chickens and sheep,

Some believe it might be a python which escaped from a circus at Ann Arbor, Mich., about 20 miles from its present haunts, last year.

The question was how a tropical snake could survive a Michigan winter.

Yesterday, a posse was led by Postmaster William T. Williams, with Davey Crockett, Ypsilanti conservation officer, acting as a guide. Sgt. Robert Winnick of the county sheriff's office went along "just to be sure nobody got hurt."

Dunkirk, New York, *Evening Observer*
June 22, 1949

Offers $500 for Snake of Salem

Detroit, June 23—(UP)—Jack Friel, manager of the Michigan State Fair, today offered a $500 reward for the capture alive of the 14-foot snake said to be on the loose near Salem, Mich.

Friel said that he will give the money providing the serpent is exhibited in the fair's centennial celebration Sept. 2 to 11.

Townspeople at Salem hunted the reportedly giant snake last Sunday but could not find it in the marshes near there.

Traverse City, Michigan, *Record-Eagle*
June 23, 1949

Michigan Posse Hunts Big Snake

Lansing, Mich.—A 10-man posse armed with guns and clubs set out today in search of "Sin", a seven-foot boa constrictor that has haunted a sparsely-populated rural area south of Lansing for the last two months.

Two burly gravel pit employees fled in terror yesterday when they saw a huge snake coiled around the frame of an old abandoned truck.

Floyd Ackerman and Richard Fuller were working at a lonely gravel pit in the vicinity where Sin escaped from a carnival. They said they saw the old truck frame in a clump of brush and started to walk toward it when a snake "as big around as a man" reared its head that was "as big as a pie plate."

Elyria, Ohio, *Chronicle Telegram*
July 11, 1951

Minnesota

1. Moorhead, Clay County
2. Brainerd, Crow Wing County

A live 15-foot boa constrictor was pulled out of a strawstack near Moorhead, Minn., a few days ago. It is thought to have escaped from a circus last summer.

Wellsboro, Pennsylvania, *Agitator*
May 8, 1912

Huge Snake Reported Seen
at Farm South of Brainerd

Myth or monster? A 22-foot python or an overgrown common variety of reptile?

Those are the questions which face authorities today as
reports of a monstrous reptile wriggling its laborious way northward emanate from the Daggett Brook community.

First sighted near Sauk Rapids, Sartell and Little Falls, the
22-foot snake with a head "the size of a stove pipe" has now been
reported on the Carl Nelson farm home 16 miles southeast of
Brainerd in Daggett Brook township.

A Brainerd policeman investigated at the Nelson farm home
yesterday and saw no reptile. "There were impressions among
the needles of the pines and in grass which could have been made
by a very large snake, however," the patrolman stated, "I could
find no evidence that the snake had crossed any of the roads which
surround the Nelson place."

Mr. Nelson reported the appearance of the reptile to Crow
Wing county and Brainerd city officials in this fashion:

"My daughter, Mildred, 15, was going after the cows one
afternoon last week when she saw the snake in a slough in the
middle of the pasture. The size of the reptile frightened her and
she raced home to tell me.

"I picked up a rifle and mounted a horse. Just before reaching the pasture, the horse began to get skittish and did not want
to go any further. Then I saw the snake lying under some pines,
about 20 feet from the slough where Mildred had said she saw it.
I didn't wait to see how long it was for its head raised three feet
in the air, was a big around as a stove pipe!"

Later, two other members of the Nelson family claimed to
have seen the huge reptile. It was then that Mr. Nelson telephoned
Brainerd to inform the sheriff's office and policemen.

County authorities had heard rumors that a carnival had lost
huge reptile near St. Cloud. A check with police at St. Cloud did
not confirm this report, however. Rumors near Sartell said that
carnival had a 22-foot python and that a reward was being
offered for its capture and return.

One report of the sighting of the snake was heard from near
Ramey. Others have come from points in a direct line from Sauk

Rapids to the Nelson home, lending credence to the belief that the same reptile has been sighted by all parties.

A University of Minnesota snake authority is reported to have stated that a python—if such the reptile is—would not survive long in cold weather.

Brainerd, Minnesota, *Daily Dispatch*
September 9, 1941

Mississippi

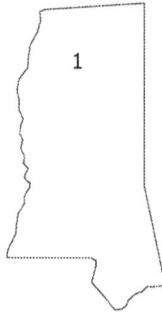

1. Yalobusha County

—There are rumors of a monster snake having been seen on Yalobusha, a few miles above this place. He is estimated to be 18 feet long, makes a mark 8 inches wide, and a coil as large as the largest wagon wheel.—Yazoo Valley *Flag*.

Atlanta, Georgia, *Daily Constitution*
August 17, 1878

Missouri

```
        1      6
    7         8
      5
           3

      4
        9
               2
```

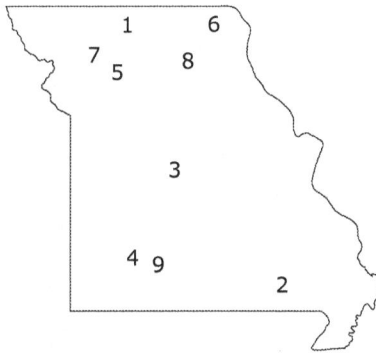

1. Mercer County
2. Carter County
3. Clarksburg, Moniteau County
4. Springfield, Greene County
5. Chillicothe, Livingston County
6. Gorin, Scotland County
7. Lock Springs, Daviess County
8. Macon County
9. Cedar Gap Lake, Webster County

In 2003, several "big snake" stories from Cedar Gap Lake were reported to the Webster County *Citizen*. The stories ranged over several decades, with several in the 1950s. They may be worth some attention, though there don't appear to be any recent sightings.

A farmer in Mercer county, Mo., killed a viper fifteen feet long. His snakeship was looking leisurely over the fence at the man plowing, and he shot him.

Indiana, Pennsylvania, *Progress*
August 31, 1871

A Wild Snake Story

A Carter county, Mo., correspondent writes:
While riding through the woods yesterday on the forks of Little Sac and Big Sac, in this county, I saw stretched across the road what I supposed to be a piece of burnt timber, but was somewhat surprised to find that I could not get my horse to approach nearer than twenty-five steps to the object. But imagine my surprise when, on closer examination, I discovered it was in motion, and proved to be a huge snake, full twenty feet long, and apparently twenty inches in circumference.

Not being able to get my horse nearer than twenty paces, I dismounted and picked up a couple of stones, for I was afraid to attack his snakeship with a club, I then remounted my horse and hurled the stones at him, supposing he would retreat; but he seemed to demur at this proceeding, and turned back in the road just in front of me, raised its head some three feet perfectly erect, as if to say, "How is this for high?" I began to think seriously about retreating, feeling that I had met with more unpleasant individual than I at first anticipated, when he again lowered his head, and began moving slowly in the direction whence he come.

I again dismounted, gathered some stones, and commenced throwing them at him, but by this time I had become excited, or something else, and could not hit the monster. He seemed to pay no attention to my throwing until he reached the hazel bushes. I then mounted my horse and urged him forward to where I could look under the bushes, there I discovered his snakeship in a position that convinced me if I intruded further he would let me know that had a tail to unfold.

So I concluded that I would excuse this narrative and go on my way rejoicing that I had not been caught in that coil. Now, this may seem to your readers a big snake story, but I am willing to be qualified as to its truth. It was of a deep blue color, with a white ring around its neck. Another party says that he saw the same snake two years ago and was afraid to attack it, and was also ashamed to tell it, fearing it would impair his veracity, but did reveal it to his brother-in-law at the time.

Cambridge, Ohio, *Jeffersonian*
December 5, 1872

A Monster Snake Killed.

One of the largest snakes we have heard of in this part of the country was killed on the farm belonging to Capt. Streby, in this county, about 3 1/2 milts north of Clarksburg, on the Pisgah road, on Thursday of last week. It was killed by Mr. G. W. Streby, who states that it was a bull snake, measuring 12 feet 2 1/2 inches in length, 3 7/8 inches in diameter, and weighed about 180 pounds. It was found by some boys who were at work in the woods cutting timber. They at once gave the alarm which brought Mr. Streby to the spot, who after considerable hammering, succeeded in killing the monster. They seem to think that the mate to this monstrous reptile is also in that vicinity, as they heard it blowing on the evening of July 3rd; the noise seemed to be about six rod's distant, but their search for it proved unsuccessful. They are still on the look-out for it.—Boonville *Eagle*.

Sedalia, Missouri, *Daily Democrat*
July 15, 1875

The Boss Snake Story
Terrible Adventure of a Missouri Farmer and His Cow.

From the Springfield (Mo.) *News*.
Last Monday, Mr. Joel Haden, of the south part of the county, wandered down on the bluffs to James River, searching for a lost cow. He finally discovered her standing in a bare, open place where she had passed through a narrow crevice between the rocks, there being no outlet. She had been gone several days and seemed

to be in a nearly starved condition, seeming not to be able to find her way out. Seeing her condition, Mr. Haden boldly started through the opening to drive her out. He had proceeded but a few feet when he felt a soft, cold substance strike him, and before he had collected his thoughts a large black snake was coiling its slimy folds around him. Almost paralyzed, he tried to flee, but it held him fast in the crevice, its tail being coiled around a shrub several feet up the side of the bluff. It began then to tighten its folds around him. Great drops of sweat stood on his brow and his hands became useless. Before his face the snake raised it head, flushed out its forked tongue at him, and he closed his eyes, expecting the end. Then his scattered thoughts collected. In his pocket he had a great, strong knife, and without opening his eyes he pushed his hands down through its folds, quickly opened it and gave a slash, but the snake only drew the closer around him till breathing became difficult. He opened his eyes again and there was that awful head almost against his face. As quick as lightning he seized the snake with one hand close at its head and with the other he made dexterous slash and cut the head off about ten inches from the body when the folds dropped from around him and he was again free. He sat down and took a rest. Then he stretched the snake out and found that it measured thirteen feet and eleven inches. In the ledge above he found its nest, with thirty-seven eggs in it. It was thus explained why the cow was afraid to come out.

Lincoln, Nebraska, *Daily Nebraska State Journal*
September 20, 1883

The worm's in and the worm's out,
Still leaving the people in doubt.
Whether the snake that made the track
Was going south or coming back.

The above beautiful and suggestive lines are frequently quoted by political speakers in besmearing each other when they have exhausted their stock of epithets. But as I'm not a candidate, nor never expect to be, I will try to tell a straightforward, unvarnished snake story as I understand it, and also what will, in all probability, happen soon after corn planting time.

Almost every spring for forty years or more, and just after corn planting time, the track or trail of a reptile of huge dimensions appears in the roads and soft ground about the rocky ledge or bluff near the old Gillasper-Ulmer mill, on Grand river, about six miles north of Chillicothe. I'm informed that so far only three persons, at different times, have ever got sight of the monster. One of those was Caleb A. Gibbons, a pioneer now long deceased, and John F. and his son, Thos. Gillespie, still residing here.

Thomas told me that he once got to see about twelve feet of the latter end of his snakeship as he went gliding into his den in the sandrock cliff, and that it was about as big around as the coffee pot on the counter, which was about seven inches in diameter, and that he believes from the color it was a rattler. He also told me that the neighbors, by agreement, met once near by and quarried into the ledge and dug out and killed 700 snakes of different varieties.

Those who have seen the tracks of the big snake describe them as being as broad as the track of a common ox sled, which is generally about five inches wide.

It is supposed by those that speak of the matter that the big snake only leaves his den once a year, and gobbles up a few hogs and returns to his solitude out of harm's way. I'm not a snake fancier, nor have I been since I first read the story of the old portent beguiling Adam's wife with a crab apple, as there was no improved fruit at that date. Might it not be that this is the same old serpent; and if not so, why should he take up his abode in civilized Poosey?

And now I'm reliably informed that one of our most energetic and astute lawyers has taken it into his head, or somewhere else, to capture the aforesaid reptile and fetch him home alive, to all of which I say, so be it. I wouldn't know how to do the job, but my friend is a science hunter and will no doubt succeed. He is going into camp near by, will "lie in wait" for him, and if that is the programme his snakeship had as well surrender unconditionally and save much bodily exercise, for our legal friend is violently in earnest and his keen black eyes sparkle and fairly snap with the fire and enthusiasm there is in him.

It is said by naturalists that serpents can charm und capture birds and animals by gazing intently into the eyes of their victims.

If ever the big snake tries his art on our friend his tail will fly up,
see if it don't. The big snake may be an ugly customer, but he'll
have an ugly one to deal with.

I would like to be present and witness the contest, and render
assistance if necessary. I know our legal friend has had republican
proclivities, but as "I hate republicanism less and snakes more,"
I would not stand idly by and let the snake eat up my friend. I
should lend him a friendly helping hand, if I could do so at long
range. X.

Chillicothe, Missouri, *Morning Constitution*
May 4, 1890

A Monster Snake
It Has Been in the Neighborhood of Gorin Four Years.

A snake that measures from twelve to fifteen feet has for four
years past been more or less of a nuisance to the people of Gorin,
Mo. People whose veracity could not be doubted have told wonderful stories of this marvelous snake. Every spring some one runs
across it. A year ago Mr. Rodney Lease crept up to its den and
watched for an opportunity to shoot it, but became so fascinated
with the sight of his snakeship that he forgot to shoot until it had
crept back into the dense undergrowth.

Every person who has ever seen the snake gives the same
description of it. It is black hooded head, at least fifteen feet
in length, and as big round as a telegraph pole. Recently, as
Mr. William Gilmore came along the track towards Gorin, he
met it stretched out its length on the bridge; he thought to get
close to it and either kill it or at least, get accurate measurements of it by counting the ties it was lying on, but as soon as
his presence became known to the monster it quickly coiled
itself, and the sight so frightened Mr. Gilmore that he sprang
backward down the embankment and lost no time in getting
to town. The dimensions he gave of it are too large to report:
however, his word is as good as gold, and a party is being made
up in town to try and capture this nameless species of the reptile family. Taking all reports into consideration, some believe
it to be an escaped boa constrictor from some show, but its

head is different from any known snake's, resembling that of a dragon.

Perry, Iowa, *Daily Chief*
September 11, 1896

Monster Snake Killed
It was Sixteen Feet Long and Terrorized the District

Farmers in the vicinity of Lock Springs, Mo., have been greatly harassed through depredations on their chickens and pigs, and the mystery was not solved until the other day, when Newton McCrary started on the trail of what appeared to be a monster snake.

He traveled a distance of two miles, when he came to the banks of the Grand river, where it appeared the reptile entered the water. A search of the vicinity later on rewarded McCrary for his persistence. Apparently asleep, after having dispatched several full grown chickens, lay a reptile of such monstrous proportions that the man was transfixed with fear. His courage returning, McCrary sent a charge of buckshot into the head of the reptile and precipitately fled.

The contortions of the snake as it lashed its tail and body against the ground and trees added to the fears of the now thoroughly terrified man. Summoning the assistance of neighbors, McCrary cautiously led back an armed party, when, to his satisfaction, the life of the reptile was found to be extinct. It measured 16 feet, and the body was as large as an ordinary stovepipe. The species of the reptile is not known, although it looks very much like a python. Early in the spring it was seen several miles farther up the Grand river, but reports of the terrified spectators were not generally credited.—St. Louis *Globe-Democrat*.

Lima, Ohio, *Times Democrat*
August 11, 1897

Saw A Giant Snake

Macon, Mo., July 22—News has reached here that the monster snake, reported to have been seen in eastern Macon County at intervals for the last thirty years, bobbed up again recently on the farm of Frank Morris about five miles northwest of Clarence. The story is that while driving his plow across the field last week near the end of the furrow, Mr. Morris' team began to rear and plunge. Looking ahead to see what the matter was Morris observed the big snake slowly moving off towards the woods. He didn't try to stop it, as his calculation of the reptile's size was "about thirty feet long and as big around as a stove pipe."

Mr. Morris' estimate of the snake's size is corroborated by Newton Pierce who lives on an adjoining farm, and who also saw it. Pierce's father, now dead, reported having seen a black or blue snake closely fitting the size indicated, so there must be one whale of a snake over in that neighborhood.

A telephone inquiry to the Clarence Courier regarding the big snake worrying the farmers up north of that town and whether its size was not a bit exaggerated, brought this answer from the editor.

This monster snake has been bobbing up for thirty years or more according to well-authenticated reports. The men who claim to have seen it are sensible and level headed, and they have undoubtedly run across something far out of the ordinary in the reptilian kingdom. Up to the north where the big snake was seen are some tall cliffs honeycombed with fox holes and hiding places for wild animals. It is supposed the snake spends his winter in one of these holes and that when hot weather comes he travels about to bask in the sunshine.

The Macon Hunters Club will probably investigate the story of the big snake and organize an expedition to capture it. Otho F. Mathews has volunteered to lead the expedition.

Chillicothe, Missouri, *Chillico Constitution Tribune*
July 23, 1924

Montana

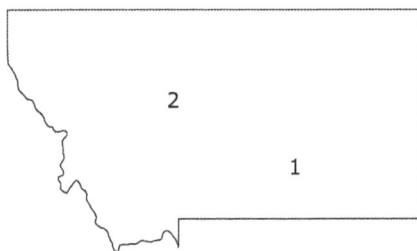

1. Laurel, Yellowstone County
2. Cascade, Cascade County

Mammoth Snake Captured Alive

Butte, Mont.—The story of an extraordinary battle between two sheepmen and a mammoth snake, coming from Laurel, a remote section of Montana, in the Little Pryor mountains, has caused great excitement among naturalists and others. The huge reptile, declared to be larger than any ever heard of before located in this region and of a species yet unidentified, was taken alive after a thrilling hunt in a wild mountain ravine whither L. N. O'Dell and J. W. Vaught had been guided by a number of terror-stricken Indians whose superstitious fright sent them fleeing from the neighborhood long before the actual capture of the snake was accomplished. The reptile is 18 feet in length and weighs 200 pounds. Around the body it measures more than eighteen inches.

At different times in recent years, O'Dell had heard from the Indians of the strange monster that made its home in a wild canyon a few miles from Laurel, but until three braves returned terror-stricken recently from the vicinity, he took it for granted that the story had its birth in the superstitious imagination of the redskins. He then became impressed with the remarkable tale and, after enlisting the aid of Vaught, started forth with a party of Indians as guides.

At the entrance to the canyon the swarthy guides deserted precipitately and the two men ventured into the narrow ravine, where a large hole entering near the top of the ledge covered with slippery shale rock and innumerable trails leading in all directions from its mouth gave unmistakable evidence that the opening was a lair. Cutting a square hole several feet back from the entrance to this wild abode, they discovered coils of a monster even larger than they had been led to expect.

Stirred to wrath by the disturbance, the snake began to make a great noise, and the few remaining redskins, that had retired thirty or forty rods, scampered off on their fleet-footed ponies at a gallop. As the monster started to leave the entrance, O'Dell threw a gunny sack over its head and clasped his arms around its throat. Then, one of the weirdest struggles ever recorded in Montana history was on. Back and forth over the narrow ledge, the huge snake writhed and twisted, fairly tightening itself about O'Dell's body until he was lifted from the ground.

Vaught seized the snake by the tail and forced it to release its hold on O'Dell. The strange contest continued upon the yielding floor of the little ravine for almost three-quarters of an hour before the two men succeeded in tiring the snake, when they bundled it into a sack and took it alive, struggling, to Laurel, as living proof of their strange story.

The reptile is marked with [—] dark mahogany spots, outlined in a lighter color, and extending across its back. Its method of killing prey is apparently by constriction, but its spots are not the shape and color of a boa, nor is it an anaconda, as its body is much too large.

O'Dell, who has had a vast experience with snakes of all varieties, does not believe the snake is a native of Montana, but thinks it has d[—] here from the Sierra Nevada mountains. The Indians claim the snake has been seen in the Little Pryor mountains for more than twenty years and have associated its doings to the actions of the "evil one." O'Dell does not believe another snake of its size could be found in the state, and will present the huge reptile to the Society for the Preservation of Natural History of Montana.

Elyria, Ohio, *Evening Telegram*
September 16, 1910

Proves 20-Year-Old Story
Capture of Giant Serpent Found in Cavern
Seems to Solve an Oft-Heard Yarn.

Helena, Mont.—Lon O'Dell and William Vaught and a part of Crow Indians have solved a snake story of twenty years' standing by the capture of a monster, the like of which has never positively been known to be in this locality before.

The snake was followed to a cavern on the Crow reservation. A hole was dug several feet back of the mouth of the cave to intersect it where the coils of the snake could be seen. The snake started to leave its lair, and as it did so O 'Dell threw a sack in its face which it seized. While the teeth of the snake were fastened in the sack O'Dell threw the rest of the sack over the reptile's head and clasped his hands about its throat just back of its head. The tussle continued until the snake was forced into the sack.

O'Dell is an experienced snake catcher and he believes that the snake is of a species found in the Sierra Nevada mountains. It is 20 feet long and 10 inches in circumference at the largest part of the body. It is of mahogany color with stripes of gray-gold. The Crow Indians have told of a monster snake for 20 years, but none believed them.

Sheboygan, Wisconsin, *Daily Press*
February 16, 1911

In October, 1978, a Helena woman told the Cascade Chief of Police that she almost put here car in a barrow pit after running over a huge snake on Highway 15, 2 1/2 to 3 miles south of Cascade. She stated, "It was between 20 and 30 feet long and its coils were at least three feet across. It covered my side of the freeway. It was standing, with its head up, and it was taller than the hood of my car. I tried to slow down and I'm sure I hit it or it struck at the car because it hit high on the left hand side of my car. It appeared to be a sort of a gray-white in color with a tan strip. It had a flat head that came down to a point and the head was wider than the body. The body was about six inches in diameter at its widest point and, from the way it stood and the shape of the head, it looked like a cobra. I've seen rattlesnakes, bullsnakes and cobras, and this looked like a cobra." The woman's daughter was also in the car and saw the snake. The police were unable to find a trace of the snake, though the Chief said that other people had reported seeing it. (Great Falls, Montana, *Tribune*, October 28, 1978)

Nebraska

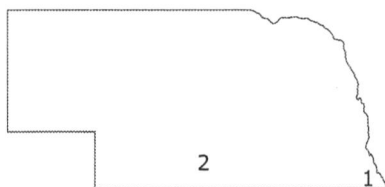

1. Stella, Richardson County
2. Holdrege, Phelps County

Citizens of Stella Terrorized by Reptile

Stella, Neb., Nov. 2.

While Mrs. J. H. Overman was doing some decorating around a grave in the cemetery last week she was terribly frightened on looking up and finding a huge snake almost beside her. She ran to town, describing the snake as being larger than a piece of stovepipe and about sixteen feet in length. Mr. Overman got about a dozen men and made a complete search of the cemetery for his snakeship, but were unable to find any trace of him

Yesterday afternoon Marvin Davis and George Smith had occasion to cross the cemetery and accidentally came across the snake, which tallied exactly with the former description. It managed to crawl through a hedge fence, however, before they could dispatch it and made its escape. A large crowd is going out from town this afternoon and will try and capture it alive if possible.

Lincoln, Nebraska, *Evening News*
November 2, 1905

Bigger And Better Snakes.

Snakes and fish are stimulating equally to the imagination. For some reason, the customary number of fish stories have not made their appearance in these parts, but it still is early in the season, and there is hope the shortage will be overcome. Fortunately, the snake comes to the rescue.

Out near Holdrege, a farmer cultivating corn in his field came across a snake easily 15 feet long and 6 inches in diameter through the thickest part of its body. It was a hot day, and the sunshine was brilliant, so, perhaps, the eyes may have been playing a few tricks. But to continue with the story, the trusty dog saw it first, and in tribute to its awesome appearance, beat a hasty retreat. Smart old dog! The farmer, lacking weapons fitted to tackle such a formidable monster, tossed a clod of dirt at it with surprising results. The snake raised itself up about two feet, spat suggestively, and then hied itself away to a neighboring plum thicket. The farmer was of the opinion that it was not a native of the community, but a fugitive from some circus.

Boy, what a thrill! Back in the good old days, the gang enjoyed the sport of dispatching water moccasins, or maybe they weren't moccasins at all. A long bamboo pole, with a sharp blue steel fishing hook tied securely to the end, and then a stealthy trek to the creek banks to spy the snakes while they were sunning themselves upon rocks and fallen logs. If the operator were skillful, a quick thrust would plant the hook firmly in the neck just below the head, and then a pull and the snake would be out on the bank, battling desperately against wooden clubs of durable weight and length. It was rather a cruel sport, but a great one, and if it did succeed in ridding the waters of poisonous snakes, it might be excused. How that gang would have liked to have run across that fifteen footer, and what it would have done to him!

Lincoln, Nebraska, *Star*
July 21, 1933

New Jersey

1. Elmer, Salem County

Saw a Spotted Serpent
A Jerseyman Knocked Down
by a Swipe of a Big Snake's Tail.

Salem, N.J., November, 5—The people of Elmer and vicinity, in this county, are much excited over an unpleasant visitor in the shape of a monster snake. A few days ago John Van Meter, a farmer, was cutting down the shrubs and suckers along the fence between his farm and that of Horace B. Shoemaker. While at work he was stuck with the tail of a large spotted snake and knocked down. When he regained his footing he made tracks for home. While fleeing he hastily observed the monster, and pronounced it as being fully twenty-two feet long and as thick as a stovepipe. When he had recovered from his fright he organized a gang of a dozen of the local residents and went in search of the huge reptile, but the monster could not be found. Fredrick Vineyard, who occupied the same farm about twenty years ago, stated he saw the same reptile when he was a tenant. It was then just as long as it is now and was exceedingly bold. Daniel Hitchner, who was a farm-hand at the same time, for years related startling stones of the "yaller cover" order about the "boay-constricter that hankered around the lower saw mill and swallowed negro babies." This is supposed to be the traditional snake, and in consequence, the whole locality is somewhat uneasy when wandering abroad. The serpent is said to make its home in a swamp just below Elmer, and parties are now out every day hunting for him.

Atlanta, Georgia, *Constitution*
November 6, 1887

New Mexico

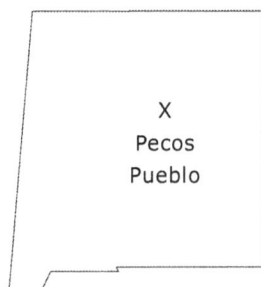

While I have no specific reports from New Mexico, there is a legend from Pecos Pueblo, where an enormous serpent deity was allegedly worshipped. (New Mexico *Daily Lobo*, February 1, 2007) This or similar tales may have influenced Willa Cather's story in *Death Comes for the Archbishop* (1927).

New York

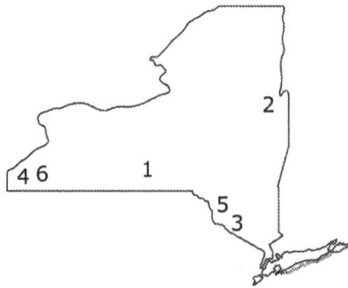

1. Spencer, Tioga County
2. Dresden, Washington County
3. Orange County
4. Fredonia, Chautauqua County
5. Liberty, Sullivan County
6. Salamanca, Cattaraugus County

A Veritable Snake Story—On Friday last, a son of Mr. David R. Gates, of this city, was fishing with two or three companions in Browning Pond, in the north part of Spencer, when, hearing a furious splashing in the water nearly half a mile distant, the party rowed their boat to the scene of action. On arriving near the spot, they saw to their astonishment a monster snake of a brown color, with its head, resembling the form of a person's two hands joined with the palms together, elevated a little distance above the water, and watching their approach. Some of the incredulous ones of the party at first supposed his snakeship to be only the dry

limb of a tree, although the water around him was in constant
motion with what subsequently proved to be eight or ten young
serpents some three feet in length, as near as an estimate could
be formed of their dimensions. The movement of the boat soon
frightened the old one, which fled in an opposite direction,
accompanied by its whole litter of snakelings. The old monster
appeared to be almost twelve in length, and performed his flight
by an up and down zig zag course, a portion of the upper half of
each soil appearing above the water. A sporting party of a dozen
or more, from this city, intend if possible to make their guns
acquainted with these rivals of the sea serpent race.—*Worcester
Transcript.*

New York *Daily-Times*
July 28, 1853

The Monster of Lake Champlain.

The Whitehall *Times*, in mentioning the discovery of the great
sea serpent in Lake Champlain, says all the gentlemen who have
seen it say it must be at least thirty or forty feet long: They all say
the sun shining on his silver like scales made the hideous length
of the monster glisten brightly. That the serpent has been seen
before we have the testimony of J. A. Parker, well known to all of
our citizens; that about eight years ago, while driving along the
public highway, two miles east of our village, he observed a large
snake eighteen or twenty feet long, and as large as a man's thigh,
emerge from the mountain recesses and move swiftly across the
field at the rate of ten miles an hour toward Jerry Collins' marsh,
and take to the water. We have no disposition to create a sensation, knowing full well that sea serpents have been too numerous
of late years, but when gentlemen who are known as men of
veracity give us the facts before mentioned, our profession as
journalists calls upon us to lay such facts before our readers. The
excitement at Dresden and the immediate locality is intense.
Parties armed and equipped are searching the marshes and bays.
Visitors are flocking to Dresden. True is it that the men at work
on the New York and Canada Railway cannot be induced to wander
off alone in the neighborhood of the monster's stamping ground,

lest their lives might be sacrificed to satiate the appetite of the great snake.

Orange City, Iowa, *Sioux County Herald*
August 15, 1873

They Saw a Monster Snake.

Newburg, N. Y., July 25.— "Johnny" Brown and "Alec" Miller, small boys, were walking along the road near the house of David Ayers, near Circleville, Orange county, when they saw what at a distance appeared to be a log lying across the roadway. As they came nearer they noticed that the surface of the object glistened in the sun and was beautifully marked with brown, yellow and white. Then it began to move.

"A snake!" shrieked the boys. Then they took to their heels. They told their story to Ayers, who was not surprised. He told the boys that he had known for two or three years that an immense serpent made its home in a cave near his place.

The snake has been seen by several persons. The boys say it was as big around as a man's body, and that it reached clear across the road, neither its head nor its tail being visible. It may be any- where from twenty-five to fifty feet long. A party of men is being organized to hunt the monster.

San Antonio, Texas, *Daily Light*
July 25, 1893

A Fredonia Snake Story

Fredonia, Aug. 7—Perhaps the largest snake ever seen in these parts was killed by Henry Berts on Arkwright hills Thursday morn- ing. Berts was hunting woodchucks in the fields back of his house, when suddenly he saw his flock of sheep start on a run. Upon investigation he found that a huge black snake had captured one of his lambs, and had coiled himself several times about the animal's body. The lamb was dead. Berts came up within 35 or 40 feet of the reptile and gave him a heavy dose of coarse shot.

The snake began to uncoil from his prey when Berts took deliberate aim and tore the head completely off with a second charge.
This had the desired effect, but the headless snake was still a lively
antagonist, and slashed his tail and body in every direction. Berts
kept at a safe distance until he had placed two more heavily loaded
shells in his gun, and then walked up to within 20 feet of his
snakeship and filled his writhing body with the contents. The
snake was of the black variety seldom seen in this section of the
country, with a bright yellow ring encircling the neck and dirty
greenish belly. It measured 11 feet 9 inches from the place where
the head was shot off to the end of the tail, and 6 3/4 inches in
diameter at the largest part of the body. This is undoubtedly the
same reptile that gave chase to a party of Cassadaga people in a
buggy about a year ago. It is by far the largest ever seen in these
parts.

Olean, New York, *Democrat*
August 8, 1893

From Liberty, New York:
    Monday afternoon one of our citizens came in from a mush-
room hunt and reported that he had seen a snake in the woods
near town, fifteen feet long. His story was not credited by everyone, but there were a few who saw some truth in it, and this
morning, armed with a shot gun, hand spikes, a wire clothes line
for a snare and a log chain to drag the animal home, if captured,
went to the woods where the serpent was seen. They could not
find it, but found a place crushed in the leaves and grasses, measuring fifteen feet, where they supposed the serpent had lain in
the sun. Failing to capture the big one, they beat no the bushes,
stirring up a large den of snakes, and succeeded in killing six black
snakes and blue racers, measuring from four to six feet in length.
It is now thought that there is a monster snake somewhere near,
and the search will be resumed tomorrow.

Middletown, New York, *Daily Argus*
August 10, 1896

Another Version of Famous Snake Story
Teamster for Frank Livery Believes His Wagon
Ran Over Big Serpent in River.

Another version of the "big snake" story which has been a matter of considerable interest here of late, is submitted today by M. D. Frank, owner of the Frank livery.

Frank says that the story was told him by one of his teamsters who has been hauling gravel from the bed of the river. Mr. Frank said that late Saturday afternoon the man was driving his team and wagon into the river opposite the foot of island park when the horses stepped over what was apparently a small log in shallow water, not far from shore. The front wheels of the wagon passed over it, but not the rear wheels. Glancing backward the driver was astonished to find that the supposed log had disappeared. The story of the monster serpent, which at least two others claim to have beheld in the river, at once flashed into the mind of the driver and turning the horses about as quickly as possible he lashed them toward the shore. He saw nothing more of what he believes to have been the big snake. Mr. Frank says that the driver asked that his name be withheld from publication since he felt that he might be made the butt of too much good natural railery if his name were disclosed.

Salamanca, New York, *Republican Press*
October 2, 1922

North Carolina

1. Bakersville, Mitchell County
2. Cleveland County

The inhabitants of Snow Creek township, near Bakersville, N. C., are terribly alarmed at the appearance of a monster snake which has been seen by several gentlemen, reliable of course, and is pronounced to be a huge specimen, about 15 feet long, two feet around the body, a head about six inches across. It is reported to be traversing the country, mounting right over tops of fences, and showing fight to all who come in its way, and has the neighborhood scared up wonderfully to such an extent that they are afraid to leave their homes. A reward of $100 alive or $50 dead has been offered for it, and a company of men raised to make an attack on his snakeship.

Fitchburg, Massachusetts, *Daily Sentinel*
October 16, 1877

It Happened in N. C. ...
William A. Shires
[column excerpt]

Out in the woods of Cleveland County there's a fearful legend of a monster snake, at least 14 feet long and as thick as a tree trunk.

According to legend, the huge reptile has roamed the woods between Thickety Mountain and First Broad River for 50 years. Quite a few residents of the Camp Creek section vow they have seen the thing.

Bob Humphries, a Camp Creek farmer, says he's been hearing about the giant snake for 33 years and about 10 years ago he saw it.

About a week ago, Humphries says, he found the trail of the phantom snake on an unpaved bottom land road behind his home.

"I'm not saying the snake I saw is the same one people have been talking about all these years, All I know is I, saw something so big that if you ran on to it, you wouldn't want to tackle it."

Burlington, North Carolina, *Daily Times-News*
July 9, 1966

Ohio

1. Fort Recovery, Mercer County
2. Clermont County
3. Newcomerstown, Tuscarawas County
4. Barnesville, Belmont County
5. Columbus and Westerville, Franklin County
6. Noble County
7. Carroll County
8. Loudonville, Ashland-Holmes Counties
9. Crawford County
10. Sandusky and Kelley's Island, Erie County
11. Wyandot County
12. Rockville, not denoted, location uncertain
13. Phillipsburg, Montgomery County
14. Sparta, Morrow County
15. Meigs County
16. Hamilton, Butler County
17. Champaign County
18. Newark, Licking County
19. Amanda, Fairfield County
20. Peninsula, Summit County
21. Salineville, Columbiana County
22. Delphos, Allen-Van Wert Counties
23. Mesopotamia, Trumbull County
24. Redtown, Athens County
25. Wayne County
26. Cincinnati, Hamilton County
27. Shelby County
28. Hambden township, Geauga County
29. Amherst, Lorain County
30. Lebanon, Warren County
31. Brown County

The American Anaconda

In the isle of Ceylon, in the East-Indies, is found the largest and most formidable serpent that has yet been discovered; it is called anaconda. Bomare says that one of these serpents was found by measure to be above 33 feet in length. It is described as devouring all the animals that come within its reach, swallowing alive the unfortunate traveller that comes in its way, and being itself excellent and delicious food when killed. There have been several accounts that this species of serpent has been seen in America. The following relation seems to leave no room to doubt, but that the anaconda is an inhabitant of both hemispheres. One of those monsters was killed on the 27th of May 1793, by a company of gentlemen who were on a hunting-party west of Fort Recovery, and by them denominated the Heterogeneo Americano. To one of these gentlemen we are indebted for the following accounts, who relates, that when killed he measured twenty-six feet seven inches and a half, and was thick in proportion. His head was green, with a large black spot in the middle; round the jaws, which were very flat, but extremely broad were great streaks, and his eyes were monstrously large, very bright and terrible. His sides were formed of streaks of bright red, green, white, purple, and pale blue, and more beautiful than can be well imagined. Down his back ran a broad stroke of olive green, twisted and waved at the edges, beside which was a narrow one of flesh colour; and on the outside of that a very broad one of bright yellow, waved and curled in various inflections. His belly was spotted all over at small distances, with large, long, and round blotches of black scales, at the edges of which stuck out large stiff bones, almost as sharp as a needle, the shape of which resembled a fish's fin. He had a streak round his neck like that of a changeable purple, and directly under his head a large white spot. When opened, there was found in him a panther, several squirrels of different species, birds, insects, and snakes of an inferior kind, all of which had been swallowed whole, and not a bone broken. As it is probable there will be many who may doubt the truth of the above, it may not be improper to inform them, that the skin is to be seen at the Philadelphia museum, where they may convince themselves of its authenticity.

Rutland, Vermont, *Rural Magazine
or, Vermont Repository*
Volume 1, June 1795

Snake Story!

Six "respectable citizens" of Clermont county, Ohio, lately made affidavit that they have seen a Snake, "full thirty feet" in length, and of proportionate thickness, in a pond near Batavia, in said county, to the great alarm of the good people of the neighborhood. What next?

Alton, Illinois, *Telegraph and Democratic Review*
January 31, 1846

The Big Snake—Again.

The following affidavit has been handed us for publication, and we give it to our readers with the single remark, that Mr. Wait is a man of respectability, and those who know him award full credit to his statement.

State of Ohio, Clermont county, ss.

Personally appeared before me, a Justice of the Peace in and for Clermont county, John Wait, and being duly sworn in deposes and says: That on the 1st day of August, 1849, between 1 and 2 o'clock P.M., I was walking down the bank of the race of Hartman's mill, on the East Fork of the Little Miami, in said county. The creek was full, and the water muddy, and, as I am informed by the miller, was between 5 and 6 feet in depth, at the point where I was. My attention was suddenly attracted by the appearance of some unusual object in the water, distant about 80 or 100 yards from me, in the mill-race, and rapidly approaching me. At first I thought it was a wild duck, but in a very short time, as it approached with great rapidity, it came on until it was directly opposite to where I was, when it suddenly dashed its head beneath the surface and disappeared. I will state that I had no idea of its being the big snake, so long said to be an occupant of the mill-dam, as I always had been up to that time an unbeliever in those reports. But now I am as firmly convinced, as I am of my own existence, that some strange and monstrous animal is there, such as I have never seen or heard of before, and I am now an old man 63 years of age, and an old pioneer of the West, and have resided in this part of Ohio for nearly 50 years, and have hunted throu'-out the West as far as Missouri. As well as I could see and observe, I will now describe it as I expect to answer to God. It was, to the best of my belief, from 30 to 35 feet in length. I judge of this from the fact that when it went down under the water, its head was exactly opposite to me and at that moment of time its tail was directly opposite a sycamore bush growing on the bank of the race, and the distance between the two points is I believe about the distance stated, 30 to 35 feet. Its body appeared to me to be about as thick as that of a man,. Its color was a dark brown or black, except a greyish color under the throat, and some white like white circles round the eyes. I could not see whether there were any scales on its body or not, nor could I distinguish anything like fins or flappers.—The motion in the water was not like that of a snake; the undulations were not up and down, but horizontal. Its head did not at all resemble that of a snake or alligator, or of any other animal I ever saw. In shape, without reference to the hair or color, it approached nearer to that of a prairie wolf than anything else I can name. There was no appearance of hair on it or its body. Its head was short and thick, unlike that of any snake I ever saw. Its eye was certainly larger than that of a horse; it appeared to me to be about as large as the mouth of a tea-cup, but the white around the eye may have deceived me, and induced me to think it larger than it was.

I have communicated the foregoing facts to many persons, and among others to Col. Thomas Kain, tavern-keeper at the corner of Main st. & 6th st., Cincinnati, and to James Perrine, Esq., of Batavia, Ohio, both of whom have been well acquainted with me for many years, and to whom I refer all those who wish to enquire further into the matter. I also refer those wishing to know my character for truth to Rev. William H. Raper, [now of Dayton,] formerly Presiding Elder in the Methodist Church; who has known me all his life.

John Wait.

Sworn to and subscribed before me, this 11th of August, 1849.

James Perrine, J. P.

Batavia, Ohio, *Clermont Courier*
August 16, 1849

Snake Meeting?

A meeting was held at Williamsburg on Monday evening, having in view the adoption of measures for the capture of the monster of Hartman's mill-pond. The proceedings will be found in another column.

Batavia, Ohio, *Clermont Courier*
August 23, 1849

The Snake Hunt.

The hunt for the Big Snake (or something else) said to inhabit Hartman's mill-pond, 8 or 10 miles East of this place, commenced on Tuesday, and had continued up to last evening without any sign of the monster being discovered. The draining of the pond, commenced yesterday, will probably take a day or two, and when that is accomplished, why, if the snake is there, perhaps he may manifest himself—perhaps not. For our own part we think it doubtful—not that we deny the existence of the veritable monster himself, for we do not choose to do so, in the face of so many reputable men who aver that they have seen him; but we do incline to the belief that he is not going to show himself when there is a crowd about, or when he may run the least risk of being captured.

The number of people on the ground daily is said to be four or five hundred. They have a 'good time of it,' judging from what we hear—some drinking, some fighting, and all amusing themselves in some way or other.

If the varmint *should* be taken before our next publication day, we shall announce the fact in an extra.

Batavia, Ohio, *Clermont Courier*
August 30, 1849

The Big Snake—Not Caught!

The hunt for the big snake was continued until Friday or Saturday, without even a sign of the monster being discovered, when the party concluded to give up the search and adjourn *sine dic*. So ends Snake Hunt No. 2.

> Batavia, Ohio, *Clermont Courier*
> September 6, 1849

An Anaconda Naturalized in Ohio.

Some two weeks ago we saw in the New Philadelphia *Democrat* some account of a big snake which had been seen in the neighborhood of Newcomerstown and Portwashington, on the P. C. & C. road. We regarded the tale as a myth: but the Democrat of the 10th inst. retained to the charge and said the big snake was no myth, and added:

"He was actually seen, and is as much of a verity as ever the great sea serpent was. A farmer from the neighborhood told a gentleman of this town that he had been seen a year ago, trying to swallow a rabbit.

His presence in that locality is accounted for by the fact that about ten years ago a menagerie, travelling through the country, had one of the wagons break down, and a young anaconda made its escape! He has now grown to full size, and may be considered a living wonder. The snake was seen by Andrew Stocker, of Salem township, in June last, while he was plowing in his field—it was as large as an ordinary stovepipe, and was standing with his head erect, and as high as his. He made a hasty retreat, but said nothing at the lime, fearing that his story would not be believed. Mr. S. is a reliable farmer, and his statement is entitled to credit. The snake has his lair near a culvert on the railroad, about a mile above Port Washington."

The *Democrat* of the 18th inst. which has just come to hand, is silent in regard to the snake. But from a gentleman of this city who passed through the locality said to be infested by the huge reptile, early yesterday morning on the rail road, we learn that the whole country in the vicinity of Portwashington and Newcomerstown was in a state of intense excitement, and that no immense concourse of people of both sexes, and of all ages, and occupations, had assembled at the former place for the purpose of having a grand hunt yesterday for the big snake.

We copy the above from the Ohio Statesman of Saturday. The people assembled in pursuance of a call for a Grand Circular Snake Hunt, by handbill printed at this office. The object was to catch the monster alive—but he wast not to be found, though his tracks thro' the swamp were numerous. We learn by a reliable gentleman that the snake was seen on Saturday morning last coming out of a hole at the roots of a large sycamore tree. The tree is hollow to the forks, where there is also a hole, and it is supposed the snake was in the top of that tree, on Friday, coolly looking down upon his pursuers. He will have to be trapped, or lassooed. Morris Creter says it is a huge "copperhead." Perhaps the ghost of Boothe.

Coshocton, Ohio, *Democrat*
August 22, 1865

A Big Snake.
A Monster Thirty Feet Long and Six Inches Thick—
He Chases Men and Boys—A Visit to his Den—
Full Description of the Reptile.

From the Barnesville (O.) *Enterprise*, 17th.
During the past year a very large snake has been seen frequently on the farm of Mrs. Joanna Kirk, in this township, about two miles from Barnesville. Mr. James Woodland, who is farming the lands on which this serpent dwells and sports, has had several adventures with it. Once in corn planting he was pursued by it, and was compelled for a few days to abandon his occupation for fear of its attacks. At another time he was cutting his outs, when he heard a rustling sound just before him in the standing grain. On looking up and ahead he saw the oats were pressed out to the right and left from a common center, and bending toward him, forming an opening about twenty feet long and three feet wide.

Presently the head of the snake came in view nearly on a level with the top of the oats, and moving directly toward him. He also had a glimpse of the reptile's body, partly elevated against the yielding stocks and partly zigzagging on the ground; and he is of the opinion that the length of the body could not be less than thirty feet, and fully ten inches across. It is certain that Mr.

Woodland dropped his cradle and 'cut dirt,' closely followed by the snake, until he was out of the field. The monster then quietly coiled himself about the shade of an alder bush to enjoy a repose; or to rejoice at the fright of the vanquished trespasser on the sanctity of its retreat. And so it remained till the long shadow of coming night warned it to depart for its rocky castle.

In raspberry time, a little colored boy, son of Albert Mabra, was in search of this fruit. Not thinking of the snake, he commenced to cross the oatsfield of Mr. Woodland. When about two-thirds through the field he happened to look back and saw, projecting fully five feet above the oats heads, what looked like a black pole, and moving right at him. All the stories of the snake flashed on his mind, and away fled the boy through the oats, with the snake close at his heels. The boy, so badly scared, outran the snake and escaped out of the field; when it gave up the pursuit.

Nobody, however, had thoroughly investigated the subject until last Saturday, when we dispatched a reporter to the scene. He succeeded in learning all about the snake, and relates his adventures as follows:

On last Saturday, after the storm, and when the sun broke through the clouds warm and shining. I determined to pay a visit to the habitat of the snake, to see if I could get a sight of it. I knew that after a thunder-storm, if the sun shone hot and bright, that the larger snakes invariably came out to sun themselves. So, armed with a revolver and a large hickory club, I started down the railroad, and crossed over to the summit of the hill, just above Mr. Cox's residence. Walking along the line fence between Mrs. Jones' and Mrs. Kirk's farms, which seems to be the pivotal point about which the perambulations of the snake are performed. I proceeded west until I came to Mr. Woodland's tobacco field. Here I observed two panels of the fence crushed down to about three rails high and across the broken fence lay what I took to be a broken log, about the size of an ordinary tier pole.

I went on a few paces further, when the sun flashed out, and the whole length of the supposed log that was in sight, shone brightly in its warm rays. I halted, when not over twenty foot from the black object, and stood still. I was satisfied it was the thing sought for. I grasped my club with a firmer hold and remained motionless for, perhaps, five minutes, when the head of the snake shot up above the top of the tobacco plants only three rows from me. The head was about a foot long, six inches wide, and of a dark-brownish color. The eyes were large or appeared to be, and as black as charcoal. They were surrounded by a white ring as large as a silver five cent piece, and gave a hideous glare to the side of the head.

To run was useless, and to attack the snake, considering its prodigious size, was too fearful an undertaking to be thought of and, as I was dressed in black clothing, I believed the reptile would take me to be a stump. So I determined to remain still & note events. Directly the head of the serpent began to rotate first to the west and then toward me, but as his snakeship kept his mouth shut, I felt sure that he did not recognize me as a man. In a little time the head bent forward and commenced to move along the tobacco row at right angles with the fence. When the head had got perhaps fifty yards distant, I looked for the black log across the fence, but it was gone. By this time the head was at least a hundred yards from me, and I resolved to make an effort to see the entire body if possible. I made a circuit through a corn field adjoining the tobacco field, and came out close to a large stump in sight of the snake's head, a little south of me. On my way I had found a couple of clapboards.

I went on the opposite side of the stump from the snake, placed one of the clapboards on the ground, and gave it a half dozen strokes with the other. At the unusual noise, the snake thrust its head and body up fully six feet above the tobacco, shot out its tongue with a clucking, hissing sound, and then darted away obliquely toward the fence. As the serpent had retreated in such haste, I grew bold and dashed down through the tobacco, and when I came in sight of the fence the snake just got to it. It poised its head on the seventh rail and slowly glided over the fence. The time consumed in passing over the fence enabled me to judge very accurately of the size of the reptile. The full length of the body, at the lowest estimate, was thirty feet, and its average diameter six inches. The track made by it in the soft tobacco grounds, was at some points two inches deep, and the radius of deviation, from a direct line, made by the bendings of the track, was slightly over two feet, where its motions were unobstructed. The bottom surface of the tracks was rough with the marks of the scales of locomotion.

Supposing that a sufficient time had been taken up, while I was making my observations and measurement, for the serpent to become quiet, I slipped over the fence and cautiously approached the ledge of rocks where report located its den. The upper side of the rock is level and smooth. So creeping to its edge and looking over, I saw the monster snake coiled into a lump at the mouth of the den. The mass, as circled up, appeared as high and as large around as the largest sized wash tub, with its head projecting from the center. Casting my eyes upward to a small rock about three feet above the larger snake, I saw another and smaller one, coiled in repose like the first.

I felt insecure in so close proximity to two such mammoths of the serpent class, and so departed, leaving them to the enjoyment of the sunshine and their home among the rocks.

The foregoing facts should admonish the people to devise some means to rid the community of these dangerous reptiles.

New Philadelphia, Ohio, *Democrat*
October 2, 1868

Snake Story.—The Columbus (Ohio) correspondent of the *Commercial* says:

At last we have a genuine sensation in a true snake story. For some time past an immense snake has been seen on the Chittenden farm, on the Werterville road, about two miles from the city. Not long since he was soon in a field, with a large rooster in his mouth. At another time he was seen in the act of crushing a cat. The sight of the monster so frightened the man that he ran away as quickly as possible, and reported the wonder, only to be laughed at. The following day two boys, in the same locality, came upon the monster, who, with darting fangs, at once gave chase, and did not slacken his pace until he had traveled half a mile. The boys told their story, and were also laughed at. Yesterday morning the boys again visited the farm, with guns and were soon rewarded with a sight of the snake, who again approached them in a warlike manner. A few well directed shot dispatched the monster, which was found to be sixteen feet in length, and proportionately large. One of the most respectable citizens of this county states that he saw and measured the snake, and assures the curious that there is no mistake as to the length and size of the monster as stated above. The skin will be stuffed and brought to this city.

Janesville, Wisconsin, *Gazette*
July 9, 1869

Large Snake.

On the 29th of October, a black snake was killed near Sharon, Noble County, which measured eleven feet and two inches. A full grown squirrel and a large toad were found in its stomach.

Coshocton, Ohio, *Age*
November 19, 1869

Snaky.—Perry township, Carroll Co., is excited about a monster 15 feet snake in that region. A correspondent intimates that it has already swallowed an old gray horse, a lot of children, and a mowing machine.

New Philadelphia, Ohio, *Democrat*
July 28, 1871

The Big Snake Killed

The Ashland *Times* learns from various sources the following particulars in regard to the slaughter of the big snake near Loudonville. "On Monday of last week Mr. Deyarmon and Mr. Gilbert, residents of Loudonville, went out hunting near that place. After perambulating over a good deal of ground, they discovered a monster snake, and at once commenced an attack. Six loads of shot were lodged in the huge carcass of the snake. It became furious and while lashing its immense body Mr. D. approached too near, was struck by the reptile und knocked down. He escaped, however, without serious injury, and the snake, after writhing in agony a few moments longer, expired. It measured twenty feet and three inches, and is supposed to be one of the number that has infested the vicinity of Jeromeville for years."

What have the women and children in that neighborhood been doing, that should make it necessary for the Times to punish them with such a fearful story. Only "one of a number." Horrible!

Elyria, Ohio, *Independent Democrat*
August 21, 1872

A Big Snake Story

This is from the Barnesville, (Ohio), *Enterprise*: A few days since while Joseph Selby was gathering raspberries, he came to what he supposed to be a log, and being somewhat tired he sat down upon it to take a rest, when much to his surprise, he commenced moving down the hill. He was so much frightened that he did not know for some time what was the propelling power, but when he recovered himself he found that it was a monster snake carrying him upon its back. He supposed it was from fifty to sixty feet long and as thick as his body. He fell off during the journey, and the snake continued down into the hollow. This snake has been seen by various parties for several years past, mostly in the raspberry season, but nobody has enjoyed such an intimate acquaintance with the monster as Mr. Selby. We would advise everybody to be on the lookout for this reptilian monster during the raspberry season.

Lancaster, Pennsylvania, *Intelligencer*
August 8, 1873

A Crawford County Snake Story

Recently Mr. George W. Churchill, who lives in or on the edge of Dallis township, in this county, declared that he saw a monster snake in the woods where he was at work; he put the snake at ten or twelve feet. Robert took up the matter and made it two snakes each fifteen feet, and represented George as greatly scared. George declares he was not scared at all, but says, as at first, there was only one snake from 10 to 12 feet long.

On the 4th of July, Old Holmes, of this county, the notorious hunter known to all the sports around here, while hunting in Dallas, strayed into Marion, about the line of Grand Prerie and Scott townships. He had with him his small undersized setter, Rose, and was taking it leisurely when his attention was arrested by the queer action of the dog, which, while it seemed to point, also evidently trembled with terror. To find what was the matter he signed to Rose to flush the bird, but she took no notice of him and cowered and trembled and whined. On examining closely, he saw about four feet from his little setter a large black snake, lying in a coil which he described like a big house-moving rope with its head erect in the center and eyeing the dog. It did not take him long to empty both barrels into that snake's head; he said the thing seemed to uncoil like a streak of lightning; there was a cloud of dust and leaves, and at the edge, his little Rose lying as if he had shot her. Without thinking of anything but his pet, he picked her up and examined her, found he had not hit her; in fact he had not seen how he could, but in big confusion he carried her to some water near by, where he sprinkled and bathed her and finally bro't her too. He was so concerned for his little curly pet that he did not think of the snake until an hour afterward, when he went to the place, and there stretched out as straight as a fence rail, lay a blue racer which measured thirteen feet four inches.

Seth says he would have skinned it and brought home the blue, but it was torn all to pieces, and his little Rose seemed to threaten a return of her fit when he handled the snake. He took her to a neighboring house, determined to return and get the skin but when he got back some hogs had already badly mutilated the carcass, and partly devoured it. So it is probable George W. Churchill did see a monster snake after all, although he has been much laughed at.

Elyria, Ohio, *Independent Democrat*
August 12, 1874.

A Big Snake Story

The Ashtabula (Ohio) *Telegraph* relates the following: "Mr. Thomas Manning, who is employed as engineer at the Lake Shore and Michigan Southern water-works, claims to have seen a monster snake one day last week, on the flats below the Lake Shore and Michigan Southern Railway bridge. Mr. Manning was walking through some high grass in this vicinity, when he suddenly came upon an object that resembled a stovepipe in size. He stooped to pick it up, when the object moved off and he discovered it to be a snake not less than twenty feet long. Manning did not stop to make any further investigations, but made tracks in an opposite direction as fast as his legs could carry him. He reported what he saw at the depot, and several men, armed with guns, went to the flats where the snake had been seen, but did not succeed in finding the object of their search."

New York, New York, *Times*
July 24, 1876

A Big Snake-Skin

The Sandusky (Ohio) *Register* says: "While some men were mowing a field on a farm of L. S. Beecher, Esq., on the Columbus avenue road, in Perkins Township, the other day, they found a snake-skin that had recently been shed, and which measured 9 feet in length, and, when spread out, was 19 inches across the middle. The snake that shed the skin must have been about 6 1/2 inches in diameter in the largest part. A portion of the skin was brought to the Register office by Mr. Thaddeus Lorch, who with others vouches for the measurements above given. It is supposed that the snake that vacated the skin is one of two reptiles that escaped a couple of years ago from a traveling menagerie at Milan. The skin is marked with large spots, and is thought to have belonged to a boa constrictor."

New York, New York, *Times*
August 8, 1878

The Sea-Serpent's Rival
Chasing a Monster Snake in Ohio
A Reptile Thirty Feet Long And Probably Fifty Years Old—
A Story Western Newspaper Readers Are Asked To Believe.

From the Cincinnati *Commercial*, Aug. 11.

In 1847 great excitement was created in the northern part of Clermont County by the announcement of old and reliable citizens, supported by their affidavits, that they had soon a snake of enormous size in Hartman's mill-pond. This pond is three and one-half miles above Williamsburg, on the east fork of the Little

Miami River. The excitement gradually spread, until public attention all over the country was attracted to the matter. A "snake-hunt" party was organized at Batavia, of which Col. Howard and Capt. J. A. Penn were the leaders, to capture the monster. A day was fixed, and at the appointed time an immense crowd, armed with guns, pitch-forks, corn-knives, clubs, and almost every other conceivable weapon, gathered from all parts of the county at Hartman's mill. After organizing and distributing the party along the sides of the mill-pond, the leaders ordered an opening to be made in the dam, which was done. The water was drawn off, but no serpent made his appearance. However, a large cavern was discovered under a high projecting ledge of stone, the extent of which they were unable to determine, but it was as large as a hogshead at the mouth, and evidently extended a long distance under the hill. Hartman was unwilling to let the opening of his dam remain unless the snake-hunters would purchase the property, which they declined to do, and as no signs of the snake were discovered, the party toward night disbanded and went home. The opening in the dam was closed, and the water rose again, covering up the mouth of the cavern. The excitement gradually subsided, being occasionally revived in the following year by reports of the great snake being seen occasionally in that neighborhood. Mr. Hartman was never able afterward to do much with his mill, as the people—even those who professed disbelief in the existence of the snake—would not go to the mill. Land in the neighborhood depreciated. A camp-meeting which had been held near there, and which before was attended yearly by many hundreds, ceased to draw, and died out. The locality seemed to be avoided. Nothing had been seen or heard about the Hartman snake for several years until yesterday. Col. J. A. Penn was going into one of the northern townships to attend a lawsuit before a Squire, and when riding along the road within half a mile of this same mill-pond, saw something on the ground, about 100 yards ahead of him, emerging from a field on the northern side of the road. Spurting his horse forward, he rode within 25 yards of the object, and discovered that it was an immense snake, not less than 30 feet long, and as large around the body as a beer keg. It carried its head about five feet high, and was moving at the rate of 10 miles an hour.

The Colonel immediately gave the alarm in the neighborhood, and in a short time there were 50 men on foot, in buggies, and on horseback in pursuit. The greatest terror prevailed among the women and children. The excitement spread like a prairie fire through the neighboring country. A telegram from Williamsburg to the Batavians put the town in the greatest excitement and confusion. The snake was going a southerly course, and was likely to pass near the town, if he did not change his route. Everybody turned out with whatever weapon he could get. Most of the people started in the direction of Afton, and on reaching there found Col. Penn, whose horse was run down, just mounting a fresh steed. He had come in sight of the snake two or three times and fired at it with his revolver, but without any apparent effect. The snake had passed about a quarter of an hour before, running directly through the yard of Hiram Sweet, and carrying off a 2-year-old boy who was sleeping in the yard. Mr. Penn, reinforced by the Batavia brigade, among whom were Col, Hulick, Judge Dowdney, Napoleon Bonaparte Ross, Dale Cowan, Postmaster Jameson, started in hot pursuit again. It was easy to track the monster, as he knocked down the rail fences wherever he crossed them, and his track across the fresh-plowed cornfields looked like the trail left by a huge saw-log dragged through them, while the corn was knocked down for several rods on each side of his path. The greatest consternation and havoc were occasioned by him. Teams ran away; the stock on the farms became panic-stricken, and rushed pell-mell over fences and through the fields of grain, destroying thousands of dollars' worth. People were running in every direction, terrified almost to death. Firearms were discharged to frighten the huge reptile. It seemed as if all pandemonium had been suddenly let loose. On went the pursuers and the pursued. The Ohio River hills were soon reached. The course of the snake was through Monroe Township and down Big Indian Creek. A part of the way through Monroe Township he kept right along the public road. Old man Simmons was riding in his buggy with Tom Nichols, when their horse, seeing him approaching, became frightened, upset the buggy, throwing them out into the road, and before they could get up and out of the way he ran by them, striking Nichols on the thigh with his tail as he whirled along, breaking the bone. Col. Penn at this time was but half a mile behind. At least 1,000 persons were at this time in pursuit, hallooing, yelling, and performing after the manner of Sitting Bull and his warriors. At every point on the route fresh recruits were added. Down the hills into Indian Creek went the snake. In a short time he was in sight of the Ohio River, and close behind was the gallant Col. Penn, and his brave and noisy followers. But they were a little too slow. The snake dashed into the Ohio and disappeared beneath its deep and muddy waters. It is believed that his snakeship took a downward course from this point. Capt. Morgan, whose boat came up in the evening from Cincinnati, says he saw three miles below New-Richmond what at first he took to be a large log floating in the river, but on coming nearer it disappeared, and he is satisfied now that it was the great snake. William Fitzpatrick and several other citizens of New-Richmond who were on the boat confirm this statement. It is now certain that the Clermont serpent was a reality, and that he has been chased into the Ohio River. Our people have returned, and the excitement has somewhat subsided. A purse has been made up for Hiram Sweet, whose fences and crops were badly injured by the monster, and who mourns the loss of his youngest born. Col. Penn is pretty sore yet from the effects of his Gilpin ride, but has the gratitude of the community for his timely discovery of this long-time terror, and for hurrying his departure from our territory. The people now feel easy. The fact is now established that the Clermont "sarpent" was no humbug. We hope that an effort will be made to kill him, as he will, no doubt, be often seen in his new element. It is now rendered pretty certain that the cavern under the bank of the creek discovered at the hunt in 1848 is where this big snake concealed himself. A farmer, who owns the land there, and has erected a dwelling-house, is now sinking a well, and having reached a bed of limestone, is doing a good deal of blasting with nitro-glycerine. This well is about a fourth of a mile from the mouth of the cavern, and is 50 feet deep. It is believed that the blasting of the rock in the well disturbed his snakeship, and probably frightened him out, owing to the proximity of the well to the cavern, or, perhaps, its actual connection with it. Kilby Hartman, Ezra Sly, and Mart Patten, of Jackson Township, have just come to town. They say that two or three smaller snakes have been seen to-day in the neighborhood of the mill-pond, and the excitement has not abated there.

New York, New York, *Times*
August 13, 1879

The Cleveland *Leader* prints a seemingly well authenticated story of the killing of a monster snake near Navada, Wyandott county, one day last week, and which measured 21 feet 2 1/2 inches in length and 19 1/4 inches around the largest part of the body.

Athens, Ohio, *Messenger*
August 14, 1879

The people of Williams Creek, near Rockville, Ohio, are in terror over an enormous serpent that has infested the place for several years. It is at least twenty feet long and a foot and a-half in circumference, of a dark brown color, and is supposed to have escaped from a circus that visited the town.

Hagerstown, Maryland, *Herald and Torch Light*
June 22, 1881

The Latest Snake Story.

Cincinnati *Commercial-Gazette*: Mr. John Robinson, an industrious farmer who resides a few miles from Phillipsburg, O., has just had a most thrilling adventure with the most monstrous serpent ever seen in this section. He was driving along the road near a strip of woods in his buggy, when his horse suddenly came to a halt, pricked up his ears and stood stone still, gazing at a large tree just in front of it along the roadside. Mr. Robinson endeavored to persuade the animal to move on, but all his urging and whipping had no effect on the beast, which stood in the road gazing intently in the direction of the tree. As Mr. Robinson sat in the vehicle, pondering on what course to pursue, he suddenly heard a rustling in the branches of the tree, and to his utter astonishment, he saw a huge serpent uncoiling itself from one of the limbs and dart down to the ground a short distance from him. Mr. Robinson declares the serpent was of an entirely unknown species, and was at least 18 feet in length and fully 10 inches in diameter. It had a monstrous flat head and darted its fangs out in a savage manner, and its eyes glowed like coals of fire. When it touched the ground it started toward the horse, and the poor beast was so frightened that it became almost uncontrollable. It trembled and shook like a leaf, and reared and plunged around at a fearful rate, but the huge serpent still approached the frightened animal, and seemed determined to dart its poisonous fangs into it. When the serpent approached to within ten feet of the buggy, and just as Mr. Robinson was preparing to leap from his seat, the monster turned around, stood still awhile, then slowly glided away through the dense forest.

Mr. Robinson says it was the most frightful looking monster ha ever beheld, and he will long remember the terrible fright he received. His horse did not recover from the nervous shock for several hours, and he is thankful the reptile let him off so easily. It is supposed to have been a snake which has escaped from some menagerie.

Perry, Iowa, *Chief*
August 3, 1883

A Big Reptile.

Westerville, O. Aug. 4—A big snake is in Johnson's woods, five miles north of town and is striking terror among children black berrying in that vicinity. When driven out it was seen stretched across the road, 11 feet wide, the head and tail being concealed in the grass on either side, hence the reptile must be 12 to l5 feet long. Its color is black.

Oshkosh, Wisconsin, *Daily Northwestern*
August 4, 1886

A Big Snake

Sparta, O., July 17.—An engineer on the B. & O. announced at Marengo a few days ago that while running from Peerless to that place he saw a monster snake beside the track. The reptile appeared to be about twenty feet long and was of a shining black color. A party set out to hunt the snake and after a time found its trail, which was remarkably plain and about eight inches wide. Dave Hunt of Bloomfield ran across the snake three miles west of this place and shot it through the heart. It measures seventeen foot two inches in length and is a remarkable sight.

New Philadelphia, Ohio, *Democrat*
July 21, 1887

Koger's Snake.

The snake story of Policeman George Koger was not a romance. He *did* kill a big snake, and it *was* sixteen feet long. He killed it on the Clint Feurt place, up the Columbus pike. It was not a black snake, though. It was spotted, yellow with white bars, and white belly. It was a pine snake, what the country people ordinarily call a "lazy" snake. They usually grow to be about five feet long, and are frequently seen high up in dead saplings or any small tree or bush denuded of leaves coiled around the branches and tied in knots, and suffer themselves to be killed without making any attempt to escape, apparently too indolent to move from the presence of danger. They are harmless. They have been known to bite dogs, but no evil effects followed. How Koger's snake came to grow to such enormous dimensions is a puzzle we leave will the naturalists. It was fully sixteen feet long, and about five inches in diameter. Its extreme laziness was probably favorable to growth. The policeman and his brother were out with their Flobert rifles shooting squirrels or whatever small game might come in their way. Becoming tired they both laid down under a tree and went to sleep. In due time they awoke, and were aware of a vile stench such as they had never smelled before and hope never to again. They thought it strange, for they were sure they had not smelled it before going to sleep. They began to look around, but did not look long before they espied his snakeship not more than eight feet away from them. It was apparently looking at them, with its head about a foot from the ground, and its monstrous proportions stretched out at full length. Koger's nerve was equal to the emergency, and taking aim with his Flobert shot the monster fair in the eye. It took it a long time to die, and in its horrible contortions and writhings stirred up the dead leaves and dust, and caused the dry twigs to snap and tremble. He doubts if the tail is dead yet. He was out to the scene a few days ago and says the carcass was still there, partially devoured by hogs and buzzards, and the stench was so great that there was no living in that part of the country.

> Portsmouth, Ohio, *Times*
> September 22, 1888

Of the monster snake which is said to have its den in an old stone quarry at Great Bend and of which we lately made mention in our Meigs county items, it is additionally said that Dr. F. M. Blaine while recently pruning hedges near that locality "narrowly escaped the monster's prodigious fangs by springing to the opposite side of the fence from where he was at work," The Doctor, while probably in a condition of excitation, says that the snake's head was fully six feet from the ground and that the odor from its mouth reminded him of the stench of a dissecting room.

> Athens, Ohio, *Messenger*
> October 3, 1889

A Big Story About a Big Snake.

G. W. Beckett, residing in the western part of the county, was in the city today, on business. Running across a Republican reporter Mr. Beckett said that he killed two black snakes yesterday, each measuring 8 feet in length

In the vicinity of Mr. Beckett's place there is a snake, which, by the residents, is claimed to be a monster in size. They further say that it is strong enough to carry away pigs and chickens. It has been seen only once, but owing to the fright of the man who happened to gaze upon the monster, his story must be exaggerated.

He claims the snake is twenty feet in length and has a head the size of a bushel measure.

> Hamilton, Ohio, *Daily Republican*
> July 29, 1892

Hunting a Twenty-Foot Snake.

Springfield, O., July 31.—It is reported that a monster snake, twenty feet long, has been discovered at Mutual, Champaign county, and parties have been out hunting the reptile all day.

Richwood, Ohio, *Gazette*
August 3, 1893

A Big Snake Story
A Snake That Measured over Fourteen Feet in Length,
Killed in Liberty Township.

A few days since Prof. W. E. Miller, who has been visiting friends in Liberty township, while out hunting killed a monster snake that had been terrorizing the whole community for a long time past. He crippled it at the first fire and then finished the work by shooting the snake to pieces with a shot gun.

While he had no accurate means of measuring the snake, he thinks it was about three feet longer than fence rail and a fence rail every one knows, is about eleven feet in length. If the Professor's judgment is correct this would make his snakeship about fourteen feet in length, the largest snake ever killed in Licking county.

Newark, Ohio, *Daily Advocate*
September 26, 1894.

Killed a Monster Snake.

Middletown, June 28.—An anaconda thirty-one feet long was killed at Amada, a few miles from here, Wednesday. The snake has been causing the farmers of the vicinity much trouble of late. Young pigs and lambs have disappeared and the owners thought a wolf had done the work. Last Saturday Farmer Sinkley discovered the snake in his hog pen. He shot at it but missed it. Two hunters named Huber and Fisher took up the trail and traced it to an old canal. A laborer named Peralte, with gun in hand, proceeded to drive the snake out.

In the hole he met the snake, which made a strike at him. This so frightened Peralte that he dropped his weapon and fainted. He was rescued and taken to a neighbors where he is now critically ill. Huber and Fisher returned to the village and a big posse of armed men went to the scene. The snake was smoked out of its hole and shot. The reptile originally belonged to a circus, but escaped in 1887 near this place. Its skin will be stuffed and sent to the Smithsonian institution.

> New Philadelphia, Ohio, *Democrat*
> July 4, 1895

Old Summit Excited

The northern part of Summit county is again wrought up over the discovery of a monster snake, which caused a great sensation and much terror two years ago. The rumors that it was still alive and roaming around several townships in the section south of Akron were confirmed Sunday by L. M. Kepler, who lives near Summit. He describes the snake as being more than twelve feet long and two feet in circumference and says that it journeys with unusual swiftness and with its head about two feet above the ground. All the farmers in the vicinity have armed themselves with shotguns and are laying for his royal snakeship.

> Salem, Ohio, *Daily News*
> May 30, 1896

Here's A Snake Story.

Salineville, June 17.—For several years it has been claimed that a very large snake had been seen near Rose Run, a short distance from here. Little faith was placed in the story until recently, when James Starkey, who lives near where the snake was reported to be in hiding, came almost upon the reptile. He states that the head of the snake was over a high fence, and its body seemed to be fifteen feet on the ground. Your correspondent has talked to three or four truthful citizens who declare they have seen the snake and claim Mr. Starkey's estimate of the snake is low. Their judgment is that it would measure a least eighteen feet and is probably six to eight inches thick. A hunting party is being organized which will endeavor to capture the thing alive.

Massillon, Ohio, *Independent*
June 22, 1896

A Homer Snake Story
A Monster that Has Frightened People in Northwestern Licking County.

Editor, Sunday *Advocate*—When I was a boy and lived at Homer, every summer for a number of years a snake had been seen by various citizens which was from all accounts the largest snake ever seen in the United States. This monster had for its hunting ground the region that is known as Lake Fork, being about two miles southwest of Homer. I remember well what a commotion was raised among the quiet citizens of Homer when good old Dr. Ayers, pale as death, came driving into town, his horse covered with foam, and related his story. He said while driving out on Lake Fork he saw ahead of him what appeared to be a black rope hanging from the top of a large oak tree, the same extending across the road over in a wheat field. On going under it, he discovered to his horror, that it was not a rope, but a monster snake wound around a limb at least 50 feet from the ground. There was 20 feet of the snake swaying among the branches, and in its mouth was a large gray squirrel. He said that the body was at least a foot in diameter, and he said that the snake seemed to be about 20 feet long; that when he drove under the monster, it uncoiled itself from around the limb, and descended to the ground, disappearing in the woods.

A hunting party was at once arranged and the entire region of Lake Fork was scoured but without success. No trace of the monster could be found. At an other time Mr. Archibald Dixon was cutting wheat in his field and noticing his horses scare, he saw to his horror this monster snake going through his field. He said that its head and a part of its body was in the air several feet and when it came to the fence, it glided on without touching the same, and that it went towards the big oak woods and hid, as the people supposed, in some secret cave. Other citizens have seen this snake, but very few have ever been close to it as it was fleet as a horse. Among those who can vouch to the truth of this article are, Dr. Ayers, Archibald Dixon, Shep. Fulton, John Landon and Calvin Hilbrant, of Homer, Becky Bell and W. N. Fulton, of Newark, but none ever had the experience with this monster as the writer. When I think of it, it makes the very blood chill in my veins and I wonder why my hair is not gray. Three years ago this summer I had an occasion to drive some cattle to a pasture near the big oak woods on Lake Fork. In order to get to the field, we were compelled to drive through the oak woods. When about middle way through, I noticed the cattle scare when they came to a long log and would jump over and run. When all had gotten over safely, it being a hot day, I stepped over the log, and sat down, thinking I would enjoy the breeze and singing of the birds, but all at once, I felt the log moving, and to my horror, I discovered that instead of sitting on a log, I was sitting on the snake itself. I had taken along with me a small flobert rifle but this I dropped and started to run. Regaining myself, I turned around and witnessed the greatest sight of my life. The huge monster glided through the woods with lightning speed. It carried its head at least six feet in the air, and it was soon lost in the swamp.

That is the last I saw of it, but I understand the snake has been seen again this year. Rube

Newark, Ohio, *Daily Advocate*
July 19, 1896

Kelley's Island Monster
Huge Snake Seen Again by Several Residents
Lay Stretched Clear Across the Road and
its Track was Five Inches Wide.

Some time ago the *Register* published a more or less authentic snake story from Pte. Pelee, Canada, in which a monster serpent, seen frequently for many years, was the central figure. Until yesterday our Canadian friends, headed by Adam Oper, the boss fisherman who was pursued and nearly caught by the dreadful creature, which chased him through a field of oats, held the season's snake record, but now Kelley's Island steps to the front with one that will hold its own.

The Kelley's Island snake has been reported seen on various occasions, but there was always reason to doubt the veracity of the parties who gazed upon it. Now, however, every element of doubt is dispelled for this time the fierce, open-mouthed, slimy, dark-brown denizen of the pretty island seclusions was seen by several responsible residents and citizens and their version "goes."

The big wiggler was not in action when seen, but lying in the warm sunshine, contentedly stretched across a dusty highway and the parties who were fortunate enough to run across it without being devoured had a good view—so far as there was room in the road, which is only about 600 feet wide. [sic] On either side there are weeds and underbrush and consequently the head and tail of the snake could not be seen, so no conservative estimate of its length could be made. It lay across the white dust, shining in the bright sunlight like a black log wet with dew and appeared to be very thick. It was seen about noon and a couple of hours later, when the shade of roadside trees was thrown upon it, it glided away. Then the track in the sand was measured and found to be five inches in width. Only a small portion of the well-rounded body had touched the dust as a matter of course and this gives room for the belief that the serpent's diameter may have been ten or 12 inches.

Erney Brothers, the Kelley's Island fisherman, reported as above yesterday concerning the discovery and what they say is no doubt authoritative, especially inasmuch as other persons, well known to them, and not they themselves, saw the wonderful and awe-inspiring sight.

Until the two snakes can be measured in some manner and this may never be, perhaps there will be a dispute between the Pte. Pelee and the Kelley's Island people as to which is the larger. Perhaps the only way to settle this will be to organize a snake hunt and run down the monsters by fair means or foul. Whether this will be done remains to be seen.

Sandusky, Ohio, *Register*
September 2, 1898

Monster Snakes.

While W. F. and S. J. Newhard of Warren, Ohio, and Willis and Lewis Goldener of Mesopotamia, were going up over the thirty-foot natural rock dam near the latter place one day last week they came upon a monster black snake that was devouring a baby calf. A gun was procured at once and both barrels, loaded with minie balls, were let go at once, striking his snakeship squarely under the left eye, and it was all over. When stretched out and measured he was found to be 12 feet 10 1/2 inches in length and as large around as an ordinary stovepipe. It was what is known as a timber black snake, and the largest one killed in that part of the country for many years. The farmers in the vicinity have been steadily missing their young lambs of late and they are now sure that the monster snake was responsible for their disappearance.

Racine, Wisconsin, *Daily Journal*
August 2, 1899

Monster Snake Killed.

Jacksonville, July 31.—A Polock living at Redtown, killed a monster snake one day last week which measured fifteen feet and three inches.

Athens, Ohio, *Messenger and Herald*
August 1, 1901

Big Snake
Has Frightened People of Two Townships and an Organized Hunt is Being Made.

Wooster, O., Aug. 5.—The farmers of Franklin and Wooster Townships, in Wayne Co., united Saturday in a snake hunt. Within the last two weeks a dozen boys have reported seeing a monster reptile in different parts of the two townships. Harvey Maize, a widely known horseman, was taken to the thicket where his hired man. Walter Jacot, said that, the ground and grass looked to him as though the carcass of a good-sized hog had been dragged along.

Thursday the farmers became satisfied with what the boys had told of the size of the snake, and during the afternoon a large party spent several hours trying to locate the monster. Oliver Mock says that he saw the snake and convinced his neighbors that the reports as to its size were true. Mr. Mock said that he was not thinking of the snake; but he came across a path in his meadow that looked as though something had been dragged along quite recently and he resolved to follow the trail. He got to the edge of the thicket along a small stream when he was startled by a swishing and cracking in the underbrush and on looking up saw a snake standing with head erect fully five feet, in the air. Mr. Mock says that he threw a club at the creature, when it dropped its head and made off. He declares that the snake was nearly as thick as a man's body and fully 25 feet in length.

The snake was also seen by William Jacot, father of Walter Jacot, who declared that it was as thick as his body and fully 20 feet long. As both Mr. Mock and Mr. Jacot are men whose veracity has never been questioned, the people of the community have accepted the truth of the story and on Saturday with all the weapons that they could raise united in a hunt, but the monster has not been found. It is thought that the snake is a boa constrictor escaped from some show.

Newark, Ohio, *Advocate*
August 5, 1901

Big Snake Killed.
It Had Terrorized Wayne County People for Weeks.

A special from Loudonville to the Cleveland Plain Dealer says: "There has been great excitement among the residents east of town over the reports of a monster snake. Within the past two weeks a number of boys have rushed to town and breathlessly told of seeing or being chased by an enormous snake. Little or no attention was paid to these stories at first until two responsible men reported having seen the monster, which left a path as though a huge log had been dragged along. One man said it was surely twenty-five feet long. A big party of men and boys started out to find the reptile and located it in an abandoned barn, near Millbrook.

The attack was made with shotguns and clubs and the snake was killed. It measured over twenty feet. It is supposed to have escaped from some circus, two having been in the vicinity this spring."

Massillon, Ohio, *Independent*
August 8, 1901

Snakes Attack Fisherman.
Collins Triumphed in Furious Battle Under a Railroad Bridge.

Newton Collins, living at Third and Russell Streets, Sovington, had quite an experience with a nest of snakes on the banks of the Ohio River under the Chesapeake and Ohio Bridge in Cincinnati yesterday afternoon. Collins was engaged shooting fish with a breechloading gun, when he discovered a snake. Getting closer he was warned by the rattler which made for him. He struck the reptile with the barrel of his gun, crippling it. He was attacked by another, which he shot. Thinking there were no more about the place he began a search.

He found four other smaller snakes of the same species, which he also killed. Collins, believing that he had exterminated all of the reptiles, started to leave, when he upturned a box out of which crawled a monster snake that made him move quickly. It looked like 20 feet in length, so he stated to *The Enquirer* last night, and as the snake showed fight he emptied both barrels of his gun into its head. He placed the snakes in a skiff and brought them to the Covington side. He then loaded the reptiles into a wheelbarrow, and displayed them on the streets, and finally took the bunch to Fred Shafer's place on Fifth Street. Shafer is an authority on snakes, and he pronounced the large snake a boa constructor. He measured the reptile, and found it to be about 9 feet long and about 4 inches in diameter. It had a regular row of spots, alternating black and yellow, and extending the while length of the back. How the reptile got in this vicinity is not known. It is thought it escaped from some circus in this vicinity, and made its way to the river. Collins was offered $10 for the snake.—Cincinnati *Enquirer*.

New York, New York, *Times*
July 11, 1903

Killed a Big Snake.

It is reported that Andy Lampkin, captain of the West End fire department, with the assistance of several other men, killed an immense boa constrictor on Maholm street Tuesday, after a terrific fight. His snakeship measured 13 feet in length and is supposed to have escaped from a traveling show.

Newark, Ohio, *Advocate*
August 25, 1903

Wow! A Kelley Island Snake Story That is the Limit

Here is a special from Sandusky to the Cleveland *Plain Dealer*: Considerable excitement has been aroused at Kelley Island by the discovery of a large strange snake, measuring about 10 feet long and as thick as a stovepipe. An attempt was made to smoke the reptile out of its hole, but it did not prove successful. Failing in this, men armed with shovels, picks and shotguns, arrived upon the scene. After uprooting many tons of soil they came to solid rock and had to give it up.

Sandusky, Ohio, *Star*
September 27, 1904.

Jerry Gump's Snake Story
Monster Blue Racer, Twelve Feet Long
Killed by Him East of Kirkwood Recently.

Jerry Gump, whose frequent appearances in the Mayor's court have made his name familiar to Piquads, is responsible for the first and biggest snake story of the season.

Jerry has been working on the McCracken farm east of Kirkwood, in Shelby county. The other day while he and his employer were working at early spring work they came across a monster snake which by their united efforts they finally succeeded in killing. It was of the blue racer species and measured 12 feet and 1 inch in length.

Jerry has not been before the Mayor in months and is on the water wagon. It is certain therefore that it is not an imaginary snake.

Piqua, Ohio, *Daily Call*
March 9, 1906

This Snake Yarn Is Real Thing

Chardon, O., Sept. 3.—Hambden is excited since that locality has become the abiding place of a monster black snake, which stretches the width of the roadway, being nearly fifteen feet long.

Hiram Toland, a farmer, saw a dark mass in his pasture and walked within twenty feet of it before he discovered the coils of a huge snake. He said the reptile lay over a slight depression of the ground in three coils and that the diameter of the outside coil must have been three feet. The reptile raised its head, showing its fangs.

Elyria, Ohio, *Republican*
September 9, 1909

Ohio Has Blacksnake Which Dines On Calves

A monster blacksnake that is fifteen feet long and two or three feet in diameter and eats calves, is the proud beast of Hambdon, O.

Some think it escaped from a traveling menagerie. No one has been able to get near enough to kill it.

Elyria, Ohio, *Evening Telegram*
September 18, 1909

Monster Snake Dispatched by Lively Farmer

L[—] Walters, a farmer and glaseng merchant of Newcomerstown, according to the Index, killed a black snake this week measuring nine feet eight inches from tip to tip and 12 inches in circumference four feet back of the head. And Newcomerstown is dry!

Coshocton, Ohio, *Daily Age*
July 14, 1910

"Seein' Things" in Quarry Town
Residents of the Middle Ridge are Frightened by Fictitious Monster Reptile

Amherst, Aug. 27:—Residents of the Middle Ridge have got "snake stories" down pat. They declare they can stand on a stack of bibles and swear that a monster reptile has inhabited their neighborhood. They give the snake's length anywhere from 10 to 30 feet. They say it has a flat tail, is as round as a barber pole and swallows frogs and other small animals alive.

People declare it may be the large monster which escaped from a visiting circus several years ago. Children, and even grownups, won't visit the neighborhood where the alleged snake lurks. A snake hunt may be instituted in a few days.

Elyria, Ohio, *Evening Telegram*
August 27, 1910

Cincinnati Zoo Lending Helping Hand
Reptile Seen Near Lebanon is Described at 20 Feet Long and Half a Foot in Diameter.
Has Lived on Farm for Last Seventeen Years
Men of Reputation for Veracity Claim to Have Seen Him and Had No Desire to Get Into a Mix-up—
Has Demoralized Live Stock.

Lebanon. O, Aug. 23—Excitement was at high pitch here today as farmers, Cincinnati zoo officials and dogs hunted a monstrous snake.

Oscar Bishop, farmer, who claims the reptile has demoralized his stock pens and terrorized his family, has posted the following:

"Reward of $25 for capture of huge snake, dead or alive, living on my farm and menacing the life of my family and stock".

Bishop describes the serpent as black, twenty feet long, half a foot in diameter with a large flat head.

"The last time I saw him," Bishop declared, "he crossed the road and crawled into a thicket. Another time I saw him catch and swallow a ground hog."

Bishop's stories were borne out by half a score residents, who reported they had seen the reptile at various times since 1908.

Judge J. A. Runyan, president of the Citizens National Bank, said he and a party of friends while motoring encountered the snake on a country road.

"We feared the snake would attack us and made no effort to kill it," the judge said.

David Thompson, who lives near Bishop, says he has kept a history of the snake. Ham Johnson, another neighbor, with his hounds, has been hunting the snake in the bogs on Bishop's farm.

Since the Cincinnati zoo has decided to join the hunt, the whole countryside is expected to turn out.

Dunkirk, New York, *Evening Observer*
August 28, 1925

Snake Eludes Hunters

Cincinnati, O., Sept. 24—Farmers and neighbors recently joined in a fruitless search for a giant snake, supposed to be living in a swamp near Lebanon, O. The snake, said to be as big as a stove pipe, crawled between the legs of a plow horse, according to the story, and scared the horse and plowman into fits. The snake is believed to be one that escaped from a circus 25 years ago.

Lima, Ohio, *News*
September 24, 1925

Two Fishermen Capture Big Snake in Bay Near Sandusky
Scientists and Zoological Experts, Puzzled as to Identity
of Huge Reptile, Believes it to be Boa Constrictor;

Serpent 18 Feet Long, Weighs 100 Pounds

Sandusky, O., July 22 (UP).—Scientists and zoological exports today sought to identify an 18-foot snake, brought ashore from Lake Erie by two Cincinnati, O., cement salesmen, who said they captured the gigantic serpent while fishing from a rowboat in Sandusky bay.

Harold L. Madison, curator of the Cleveland museum of natural history, viewed the snake and told the United Press he believed it was a boa constrictor.

"It is a tropical serpent," he said. "It has the marking's of a boa of the constrictor type, but I am not sure of the identification. Boa constrictors seldom grow larger than 10 or 11 feet long."

The curator said both boa constrictors and pythons could live in water, but would have to come to the surface often for air. The snake either escaped from a zoo or was deliberately placed in the lake, he declared.

Capture of the huge reptile ended reports of a "huge sea monster" in the lake. The reports were first circulated last week when fishermen told of seeing the "beast." They said it sprayed water from its mouth and tail.

Reporting the capture to authorities, Clifford Wilson and Francis Bagenstose, the salesmen, said they were en route home from the Shrine convention in Cleveland, and stopped here to fish in the bay. They rented a boat and rowed about 500 yards from the shore, they said.

"We had just cast our lines when the serpent raised its head from the water near the boat," Wilson said. "I grabbed an oar and struck it, thinking it meant to attack us. The snake rolled over in the water and lay still. We thought it was dead, and hauled it aboard the boat."

The cement salesman said they rowed ashore with their strange catch where Abe Breniser, proprietor of the boat house, helped them land the serpent. The snake revived and started to get active after it was placed in a box, he said.

A crowd of several thousand curious spectators, eager for a glimpse of the "monster" which had terrorised swimmers and fishermen near Sandusky, assembled to view the serpent.

The serpent is 18 feet long, and weighs approximately 100 pounds. The back is a dull gray, and the belly white. It has a yellow head with a black coronet. The head is 9 inches long and 6 inches wide. The snake's eyes are a dull gray, set deep in the head.

According to Madison, the two salesmen risked their lives to capture the snake. He said the serpent was big enough and strong enough to crush a horse.

Charleston, West Virginia, *Daily Mail*
July 22, 1931

'Sea Serpent' Taken in Lake Erie Tuesday
Experts Believe Huge Snake Had Escaped
From Zoo Somewhere in Ohio.

Scientists and zoological experts today sought to identify an 18-foot snake, brought ashore from Lake Erie by two Cincinnati, O., cement salesmen, who said they captured the gigantic serpent while fishing from a rented rowboat in Sandusky Bay.

Harold L. Madison, curator of the Cleveland Museum of Natural History, viewed the snake and told the United Press he believed it was a boa constrictor.

"It is a tropical serpent," he said. "It has the markings of a boa of the constrictor type, but I am not sure of the identification. Boa constrictors seldom grow larger than ten or eleven feet long."

The curator said both boa constrictors and pythons could live in water, but would have to come to the surface often for air. The snake either escaped from a zoo or was deliberately placed in the lake, he declared.

Ends Wild Stories

Capture of the huge reptile ended reports of a "huge sea monster" in the lake. The reports were first circulated last week when fishermen told of seeing the "beast." They said it sprayed water from its mouth and tail.

Reporting the capture to authorities, Clifford Wilson and Francis Bagenstose, the salesmen, said they were en route home from the Shrine convention in Cleveland, and stopped here to fish in the Bay. They rented a boat and rowed out about 500 yards from shore, they said.

"We had just cut our lines when the serpent raised its head from the water near the boat," Wilson said.

"I grabbed an oar and struck it, thinking it meant to attack us. The snake rolled over in the water and lay still. We thought it was dead, and hauled it aboard the boat."

The cement salesmen said they rowed ashore with their strange catch where Abe Breniser, proprietor of the boat house, helped them land the serpent. The snake revived and started to get active after it was placed in a box, he said.

Wilson and Bagenstose took the serpent to a Sandusky garage after the manager of their hotel would not permit them to keep it in their room

A crowd of several thousand curious spectators, eager for a glimpse of the "monster" which had terrorized swimmers and fishermen near Sandusky, assembled to view the serpent

Had Close Call

The serpent is 18 feet long, and weighs approximately 100 pounds. The back is a dull gray, and the belly pure white. It has a yellow head with a black coronet. The head is nine inches long and six inches wide. The snake's eyes are a dull gray, set deep in the head.

According to Madison, the two salesmen risked their lives to capture the snake. He said the serpent was big enough and strong enough to crush a horse.

While authorities were puzzled as to how the snake, undoubtedly a land serpent, got in the lake, they have advanced several theories. It was recalled that two boa constrictors escaped or were stolen from the Toledo Municipal Zoo several months ago.

Authorities said the snake might be one of the two boas or it might be a python believed to have escaped from an Erie Railway express train when the train was passing through Northern Ohio last spring.

Wilson claimed ownership of the serpent today, and said he probably would take it to Cincinnati with him.

"I might be able to sell it," the salesman said. "If not, I will give it to the Cincinnati Zoo. Meanwhile, I hope to find out exactly what is it."

Athens, Ohio, *Messenger*
July 22, 1931

Serpent Taken From Lake Erie
Strange Reptile of Immense Length Captured in Waters of Sandusky Bay.

Sandusky, July 22. (INS).—Lake Erie residents who chuckled over the gossip of the past week that a huge sea serpent was haunting the waters of Sandusky Bay, showed evidences of belated alarm to-day with the capture of the reported monster, an 18-foot snake.

The reptile, believed an Indian python, or *python molurus*, was taken from the lake by Clifford Wilson and Francis Bagenstose, cement company salesmen of Cincinnati. The snake was expected to be taken to Cincinnati to-day for presentation to the Cincinnati zoo.

Discovery of the serpent, the reports of which were regarded as unreliable until the snake was displayed, brought Harold L. Madison, curator of the Cleveland Museum of Natural History, to the scene.

Madison termed the reptile an Indian python, overriding an early theory that the snake was a boa constrictor.

"The markings are definite. There is a lance-shaped spot about two inches long on the Indian python's head and this appears clearly on this snake," he stated.

Speculation was rife to-day as to how the snake had gotten into the lake and from whence it came.

"The reptile is not a native of Lake Erie," Madison said, "and you can't find another within a thousand miles of it."

The curator's theory blasted an earlier belief that the serpent was one of two boa constrictors which had been stolen from the Toledo zoo several months ago.

Wilson and Bagenstose related that the snake rose up beside their boat as they were fishing in Sandusky Bay late yesterday. Wilson stated that he saw the reptile first and struck it with an oar.

The snake, he said, rolled over on its back and he hit it again before dragging it into the boat. It was believed dead until the two men took it to shore when it became active and was placed into a box.

Van Wert, Ohio, *Daily Bulletin*
July 22, 1931

Mebbe We're Wrong

Did you read that story about the sea monster captured at Sandusky, Ohio, which turned out to be an ordinary python? Unusual, wasn't it?

Now just supposing Mr. A. is a carnival or circus man whose show isn't going so good at present. He is approached by Mr. B., a publicity agent, who wants to know if A. wants to do business with him.

"Have you got any pythons or other large snakes lying around in your show?" asks the publicity man.

"Sure, I got a couple of 'em," is the reply.

"Well, lend me the sleepiest one you have for a couple of weeks. I'll secretly take it to some place far away from where you're showing. I'll get a rumor started about sea monsters, and about a week after that a confederate of mine and I will 'capture' this python out in the lake a short distance. The story will get in every newspaper in the nation. Then a little while after that you 'buy' the snake from me and go around the country billing it as the swimming python, the largest water snake in the world, the sea monster or whatever you want to call it."

Sheboygan, Wisconsin, *Press*
July 24, 1931

Huge Snake Reappears After Seven Years

Georgetown, O., Sept. 11—(AP)—After a lapse of seven years, Brown county's "big snake" is reported to have reappeared.

Seven years ago a sawmill gang, working in the sterling-tp section, sought for the "big snake" when a woman reported she saw the reptile. The search was fruitless.

Hubert Wallace, of the township, recently reported he came across a similar "big" reptile on the Mack Hesler farm. He described it as about 18 inches in circumference, with yellow stripes across the back. And so, the snake hunt is on again, with Brown countians combing the fields in search of the reptile.

Lima, Ohio, *News*
September 11, 1931

Reports Seeing Snake 15 to 18 Feet Long and Thick as His Leg

Peninsula, O.—There may be a snake hunt in Peninsula Sunday—to catch the whopping big one "as thick as your thigh."

The story has been spreading and building up into fantastic proportions around Cleveland and Akron about the big snake that Clarence Mitchell saw.

But the straight of it from Mitchell is that he saw a snake that may have been 15 to 18 feet long and as thick as his leg.

Mitchell said he saw it while hoeing the weeds from his corn ground between the old Ohio canal bed and the Cuyahoga river. He said the snake slipped into the river and "swam across and headed toward the yellow clay slip on the nose of one of those hog-backs between the gullies east of the river "

"I went over and looked at the path she made in the soft ground," he said. "Look like one auto wheel had rolled across that field.

"She was kind of dark brown all over without any marking I could notice."

Now, the small boys are whacking the bush with clubs and telling what they'll do to that ol' snake, anyway, and the women are jittery about going into the fields.

Art Huey, police chief, says there may be a hunt Sunday. Mitchell, he says, doesn't see snakes unless they're there.

Coshocton, Ohio, *Tribune*
June 20, 1944

Posses Formed For Snake Hunt

Peninsula, O.—The "Bring-em-back-alive" boys will try to capture the "Peninsula Python" tomorrow.

Mayor John Ritch announced that several volunteer posses, each in charge of a policeman, will scour the nearby swamps and hills for the huge reptile which reportedly measures 19 feet.

Posse members will carry clubs, knives, pitchforks or similar weapons, Ritch said, but only the policeman in charge may carry a gun. He added that every effort will be made to catch the snake alive.

Ritch said the best technique apparently is to try to slip a sack over the reptile's head and keep him stretched out to his full length so he will be unable to constrict. The decision to try to capture the snake was reached after experts said Pythons are non-poisonous.

The snake's tail first was discovered last Wednesday. Then yesterday, Mrs. Roy Vaughn reported he swallowed one of her chickens whole and jumped a four-foot fence when he could not get thru the woven wire.

Coshocton, Ohio, *Tribune*
June 24, 1944

Ohio Village Plans Hunt For Reptile 18 Feet Long

Akron, O., June 24. (AP)— "Bring your own guns," was the instruction of Police Chief Arthur Hewey today as he called for a snake hunt in the nearby village of Peninsula. The hunt will be held Sunday and it was called after Mrs. Ray Vaughn said she saw a reptile 18 feet long devour a full-grown chicken. Assistant Police Chief Dale Hall said he saw a snake track "made by a reptile six or seven inches in diameter." "A snake that big might kill somebody," said Chief Hewey as he advised the hunters to bring firearms.

Reno, Nevada, *Evening Gazette*
June 24, 1944.

Big Snake Still At Large Despite Search on Sunday

Peninsula, O.—The Peninsula python—also known variously as "Sarah the Snake" and "That Thing"—still was at large today.

Scores of men from Peninsula and other northern Ohio communities scoured the area adjacent to the Cuyahoga river yesterday in an effort to bring the reptile back alive.

But Python—or Sarah, as the case may be—was uncooperative.

The mass search ended abruptly at noon when the fire siren wailed three times—the prearranged signal that the snake had been sighted.

The searchers went to the described area but after a futile search Mayor John Ritch labeled the warning a false alarm.

Later, Earl Ganyard, leader of one of the posses, reported that he had discovered a new trail made by the snake since morning near the chicken yard of Mrs. Roy Vaughan who said the reptile scaled a four-foot fence after swallowing one of her chickens whole.

Experts estimated the chicken will last the snake until mid-week and Police Chief Arthur Huey said another effort to capture the snake will be made when he—or she—is sighted again.

Coshocton, Ohio, *Tribune*
June 26, 1944

Peninsula Python Is Ordered Killed

Peninsula, o., June 28. (UP)—The Peninsula Python, alias Sarah the Snake, was under sentence of death today—if and when someone is able to find it.

The sudden switch in strategy from the original plan to take the huge snake alive was announced following a conference between Mayor John Ritch, Police Chief Art Huey and assistant chiefs Dale Hall and Dud Watson.

Peninsula presently is divided into two camps, they explained, one consisting of persons who have seen the 18-foot reptile and the other of persons who doubt that such a snake really exists.

Piqua, Ohio, *Daily Call*
June 28, 1944

Claims He Saw Python in Dresden Vicinity

Bernard Crozier of Dresden may be called upon to identify the monster reptile reported prowling around Peninsula, O., because he claims to have seen the same snake three years ago.

His story is that he became so frightened when he saw the serpent "with a head as big as a horse's and a great, round body" that he fell in the Muskingum river while fishing. He was rescued by his brother and a cousin and was so scared he couldn't tell what happened.

Bernard, who was only 15 at the time, insisted the reptile chased him into the river and said its mouth was wide open as it advanced.

Coshocton, OH, *Tribune*

June 28, 1944

'Peninsula Python' Seen Near Cleveland

Cleveland (UP).—The peninsula python has been reported seen near the outskirts of Cleveland.

George, 11, and his brother, Albert Gale, 13, said they saw it near a creek where they had been swimming, but police found no trace of it.

"I saw its tail first and thought it was just an ordinary snake," Albert said. "Then more of it came to the top of the water and I hollered.

"We grabbed out clothes and ran up the cliff. When we got to the top we put on our pants and ran home."

Charleston, WV, *Daily Mail*

July 13, 1944

'Er, Sobriety?

Peninsula, O., Sept. 19, (UP)—Here we go again, folks.

Mayor John B. Hitch declared a "public emergency" following reports that the recently quiescent "Peninsula Python" had made a reappearance in the Cuyahoga river bottoms.

In a proclamation urging squirrel hunters to join the search for the snake, the mayor said that the python "is a menace to the peace and order, health and welfare and generally good reputation for veracity and sobriety of our citizens."

Piqua, Ohio, *Daily Call*

September 19, 1944

Newarkite Kills Giant Blacksnake

Clarence [McC...], 5271 First Main street, brings the prize snake story of the season to light. While picking black berries on the Charles Babb farm seven miles west of Zanesville on route 40, he saw a large blacksnake stretched out on the ground. He beat the snake to death with a club. He reports the reptile measured 14 feet long. It is the longest snake of any kind seen in that vicinity in a long time.

Newark, Ohio, *Advocate*
July 26, 1946

From Rumer (1999), on the Scioto Marsh region: "More than a hint of credibility, though, attends the stories about a giant snake accused of attacking a horse and nearly overturning a farm tractor. Sightings, real or otherwise, of the fearsome marsh snake often increased in number during wild-berry-picking season. Larry Risner tells us his father ran over 'a very large snake' while disking a muck field, and the heavy farm implement failed to sever the snake. That is the most concrete instance of snake sighting I have heard. The Risners assumed the thing had escaped from a traveling carnival and had made the marsh its home. As the stories go, this one is fairly reasonable. But assumptions about the presence of a large pythonlike snake persisted for several years."

Oklahoma

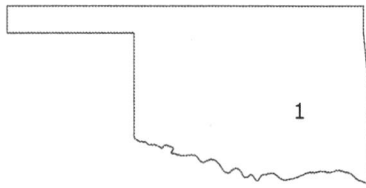

1. Holdenville, Hughes County

Six-Man Safari Seeks to Find Giant Snake as 'Long as Two Trucks, Thick as a Man'

Holdenville, Okla., Aug. 10. (U.P.)—A six-man safari pushed off into a swampy area near here Thursday, determined to track down a giant snake as "long as two pickup trucks and big around as a man."

The hunters, armed with rifles and pistols, rode horses into the forbidding area on Wewoka creek, northeast of here. They were accompanied by a "snake dog," guaranteed to "make the catch."

Reports of the huge reptile date back 15 years, according to files of the Holdenville Daily News, but today's hunt was planned after a farmer reported he spotted the snake on a road near the creek.

The farmer said he first thought the snake was a log, but changed his mind when he got a closer look.

"I jumped off my tractor and ran," he said. "When I went back, it was gone."

Four men searched the swamp Sunday. They turned up nothing but another snake story.

An Indian told the hunters he had seen the snake, which he said was "long as two pickups and big around as a man."

The snake's tracks have been seen in pastures by other farmers, including one who said it "looked like someone had dragged a log through the grass."

Tony Glass of Sasawaka, Okla., guaranteed his "snake dog" would find the snake, but if the hunters return empty handed, another party of 50 persons will trek into the swamp Sunday.

Butte, Montana, Montana *Standard*
August 11, 1950

Now Oklahomans Are Hunting
A Stray Boa—No Success!

Holdenville, Ok, Aug. 10. (UP)—Six armed hunters today called off a hunt for a boa constrictor near here because they feared too many "amateur Frank Bucks" might join in the safari and get shot.

Stalking of the legendary snake, described as" long as two pickup trucks and big around as a man," was postponed until Sunday.

The hunting party, armed with rifles, pistols and a snake dog guaranteed "to make the catch" had planned to head into the forbidden swampy area on Wewoka Creek northeast of here tomorrow.

The six hunters will be joined by some 35 members of the Holdenville Roundup Club Sunday. Many of the Roundup Club members, mounted on horses, will carry rifles.

The biggest mystery—besides the whereabouts of the snake—is what does the reptile feed on. There have been no reports of missing livestock or fowl, the usual menu for the giant boas. Principal crops in this south central Oklahoma area are corn and peanuts.

Reports of the big snake date back 15 years, according to files of The Holdenville Daily News. But the latest hunt was touched off after a farmer reported he spotted the snake lying across a road near the creek.

Bill Mullins, who lives in nearby Wetumka, said a boa constrictor was killed in that area eight years ago. He believes the latest snake also is a boa.

More skeptical residents of the area claim the "monster" is a blacksnake, a rattlesnake or "came out of a whisky bottle."

Oklahoma, which witnessed the problems of Grady, the cow in the silo, and Leo the leaping leopard, waited calmly for news of the snake which was named by one news commentator as "Slippery Sam, the Monster of the Marshes."

Galveston, Texas, *Daily News*
August 11, 1950

Giant Snake Taken Alive In Oklahoma

Wewoka, Okla., Aug. 25.—(AP)—A 26-foot snake, the likes of which have never been seen in these parts, was exhibited today by an animal trapper who said he trapped the reptile alive last night after a two-hour struggle.

The snake could not be identified immediately but was 7 1-2 inches in diameter and weighed about 250 pounds. It was dark brown, had two green stripes down its sides and diamond markings. Some said it looked like an over-grown diamondback rattlesnake.

G. W. Hall, Tonkawa, Okla., zoo keeper and animal trapper, said he captured the snake in a snare after patiently setting traps for a week and a half.

Hall released the snake in a plywood cage early today.

Hundreds of persons sleepily arrived to look at the snake after reports quickly circulated that the reptile had been caught.

The cage was 24 feet long and four feet wide. The snake did not have room to stretch out completely.

It hissed when first released, then lay quietly.

He said he could not identify the snake but "it sure is big. It isn't like anything I've ever seen before."

Hall said he captured the snake six miles north of here in Wild Horse Canyon

This hilly region of southeast Oklahoma at one time was part of one of the nation's biggest oil pools.

A "monster" first was reported in this area 15 years ago. It has been reported seen every few years since then.

Last month, a farmer reported seeing a snake "as big around as a man and as long as two pickup trucks."

Authorities had discounted the reports but several private parties have been seeking to trap the snake.

Hall said he couldn't see the snake too plainly because it was dark but said it appeared to be black with diamonds on the body.

With him when he said he caught the snake after a two hour struggle was Will Adams of Tonkawa.

"We used a pole, some rope and a net for the snare and baited it with a chicken," Hall explained.

"We caught it last night but it fought for two hours before we could get it into a canvas bag. It was so heavy the two of us couldn't carry it. We had to roll it to the car.

"There's no snake like that in the country. Don't know what it could be.

"It was too hard to measure but I'd say it was more than 14 inches around. It would take about nine steps to pace it off, I figure about 26 feet."

Hall intends to exhibit the snake.

Indiana, Pennsylvania, *Evening Gazette*
August 25, 1950

Find That Monster Snake 'Found' Near Wewoka Imported

Wewoka. Okla., Aug. 25—(AP)—Seminole County Attorney Jack Scott said today the 26-foot snake reported captured here last night was actually purchased for $450 from a Maryland firm and brought here for exhibition.

Scott said G. W. Hall, Tonkawa, Okla., zoo keeper and animal trapper, made a statement to him, Sheriff John Sandlen and Police-Chief Bill Nicholson that he wanted to capitalize on rumors in this area that a huge snake was loose.

Hall exhibited a snake today that weighed 250 pounds and measured 7 1/2 inches in diameter and 28 feet long. He said it was captured in a snare last night in Wild Horse Canyon six miles north of here after a two-hour struggle.

Scott said Hall told him he purchased the snake for $450 from a firm in Maryland. He said he saw the firm's advertisement for snakes in a magazine.

Got It At Wichita
Scott said Hall told him the following:

Last night he went to Wichita, Kas., and picked up the snake which was shipped there. He brought it to Wewoka in his automobile and at 2:30 this morning awakened a lumberyard owner and asked for lumber to build a cage.

The news quickly spread that a huge snake had been reported captured. When released in the cage, it hissed and then lay quietly until disturbed by flash bulbs of camera men.

It was dark brown, two green stripes ran down its sides and there were diamond markings on its back. Nothing of its size and description ever had been seen in these parts and it could not be identified immediately.

Followed Reports of Monster
Last month a farmer reported a snake lying on the road that was as big around as a man and as long as two pickup tracks.

Residents in this oil area of southeastern Oklahoma watched their steps carefully when treading outdoors.

Scott said he would permit Hall to continue exhibiting the snake but would allow no publicity tie-in with reports of the big snake in the area.

Authorities have discounted the snake reports but several private parties have been seeking to trap the snake.

Ada, Oklahoma, *Evening News*
August 25, 1950

Monster-Snake Tale, Plus Dead Hogs, Sets Off Some Tall Ones
Local Officers Embellish Incident with Some Real Sizeable Accounts until Simple Truth is Admitted.

County officers for a time thought Wewoka's monstrous snake took time out from his wanderings there and visited Pontotoc county.

However, they decided that this county's snake was at least 100 feet long and as big around as two men.

Friday five hogs were found dead near the side of a road three miles south of Oil Center. Part of the hogs had been eaten away.

Deputy Sheriff Darwin Cummins, who investigated the case, could find no logical reason for the hogs' deaths, so he told fellow officers, "The Wewoka snake must have killed them."

Just to make the story more vivid, another deputy sheriff, Ernest Miers, said the Johnson grass around the road "looked as if an oil tank had been dragged through it."

The story continued to grow and take on a number of interesting details—until Cummins confessed.

It seems the hogs had drowned in a recent rain, and had been discovered only Friday morning.

Ada, Oklahoma, *Weekly News*
August 31, 1950

Ontario

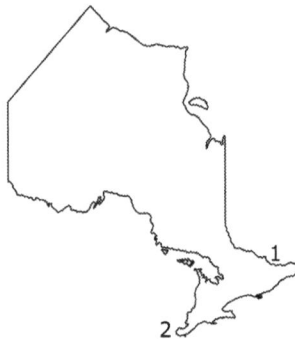

1. Ottawa
2. Point Pelee

A Very Big Snake.

A telegram from Ottawa, Canada, to the Montreal *Gazette*, dated the 16th inst., contains the following: "A young man named Pierce, who owned some wild hay at Turner's Meadow, on the Upper Ottawa, heard that some parties were stealing it while he was in the shanties. On Friday he came down to see if the reports were true. When examining what was left of the stack he saw a large snake move its head from under the stack. Pierce turned and ran away, pursued by the snake. A man named Armstrong and his son came to his assistance and killed the snake, which measured 16 feet 2 inches."

New York, New York, *Times*
March 18, 1881

Two Big Serpents
An Awful Monster on Land and Another in the Lake.
Adam Oper's Fierce Chase
Through his Rye Field on Point Pelee, Canada
Fish Nets Lashed to Pieces by a Huge Mad Denizen of the Sea—
Seen by Fishermen.

For a number of years vague stories about huge serpents have come with each recurring season from the Dominion shores, and now, at last, the existence of these fierce monsters is verified and the fact so well established that it cannot longer be questioned. They were always believed to be inhabitants of the lake in the vicinity of Point Pelee and the theory proves correct for within the past three days they have been seen there on land and in the water, under the most startling circumstances and by men whose word is reliable beyond a doubt, in fact as good as a bond.

Adam Oper, the head fisherman and also the leading farmer on the point, was the man whose lot it chanced to be to make the first discovery and he doubtless wished a thousand times that he had never made it for his experiences with the fierce, ugly, coiling thing, call it snake or what you will, was thrilling and trying and he has not recovered from the sudden shock. Mr. Oper, who, by the way, is one of the biggest and one of the jolliest men on the point, has a large field of rye near the lake on the southern shore and commenced to harvest last Tuesday. He was particularly anxious to have the shocking well done to resist the strong winds that sometimes sweep over that section and concluded therefore to do it himself. All Tuesday forenoon he shocked, exercising great care to get every sheaf just so. After the dinner hour he started for the field with his men and when nearly there all were greatly surprised to see three shocks go down in succession, as though a cyclone had struck them, when, as a matter of fact, not even a gentle breeze was stirring. They let the mystery go unsolved and proceeded with their work but after a while, when nearing the three shocks already down, Oper saw another topple very suddenly and concluded to set it up again and investigate.

He did not set it up but investigated and wished he hadn't done even that, for he came across something that made ice-cold chills play hide and seek up and down his back and great beads of sweat stand on his brow, while his hair stood up and raised his hat like porcupine quills. He picked up the capsheaf and just then he heard a peculiar grating sound and the whole mass of rye began to move. In the center a horrible head with glistening eyes and darting tongue was slowly raised and gradually a long, writhing, bluish-black body twisted itself from beneath the sheaves, reaching entirely around the fallen shock. The poor man stood like a sphinx, immovable and unmoved, transfixed by the terrible sight that met his gaze. How long this lasted he does not know, but it seemed hours, though it was more than likely but a few seconds. He was charmed but for an instant the head and sharp eyes turned. The spell was broken. Oper could move and move he did, faster than his legs had ever carried him before or ever will again. He ran like one possessed for behind him he heard the rustling of the stubbles as the rough but slippery body of the great creature glided through. For half a mile the stern chase went on, the men in an other part of the field seeing it but being powerless to help and too badly frightened to give pursuit. Finally the snake turned and moved leisurely toward the water and the panting fisherman ran a few hundred feet farther and dropped exhausted, expecting to be crushed and devoured. When he was not pounced upon he looked up and saw his pursuer in the distance and it soon disappeared in the lake. It was fully as thick as a man's thigh and Mr. Oper and his men estimate its length at between 25 and 30 feet and believe it was more nearly the latter. It carried its big flat head about the height of a man above the ground as it rushed along and looked like an ordinary snake in form, though its body looked a little flat. In the sunlight it was glossy and frightful in appearance.

Guns were secured and the shore examined but only two big trails in the sand were to be seen, marking the places where the serpent had come up from and gone back to its home in the water.

The remainder of the rye in that field has not yet been harvested and anybody can have it for the asking for Mr. Oper and the other fishermen and farmers will not go near it. The owner has offered it to the others but they do not want it.

It was on Wednesday morning shortly after sunrise that a serpent was seen in the water, this time by another party of fishermen, who lose considerable twine through it. They were sailing out to lift their second net and were still far from it when they observed an unusual commotion. Something was making a terrible stir and when they got near enough they saw what looked like coils of wire or rope a couple of feet thick twisting about and heard fearful sounds, for the water was being lashed into foam and their twine torn into very small pieces by the maddened monster, which had become entangled in the lines. They sailed away as rapidly as possible and the last they saw was an enormous black body arching itself upon the water for a moment and then plunging beneath, while a stream like a miniature waterspout shot from its nostrils high into the air. They estimated the length of the sea serpent, for that is what they saw, at 100 feet, and they had considerable time to observe it. They did not visit the spot until Thursday morning and then they found their twine destroyed beyond repair, several of the stakes even having been pulled up. A reign of terror now prevails and the fishermen are almost frightened out of their wits. It is believed that the monsters of the deep have driven all the fish away for during the past three days the catch has been remarkably light. Both yesterday and Wednesday the steamer Louise brought but a small quantity from the Canadian shores instead of her usual cargo of many tons. Last year the larger of the serpents was seen in Pigeon Bay, on the north shore of Point Pelee.

It is probably not necessary to remark that this is not a fish story but a sea serpent story and a true one. Doubting ones can have their doubts cleared away by consulting Captain D'Clute, Mate Voigt, Engineer Zanger or any other of the veracious boys on the Louise, who will verify all that has been told. They come in contact with the fishermen every day and know the ones who tell the truth and are therefore to be believed.

The larger of the serpents seen this year corresponds with that seen last year, excepting as to size, being somewhat larger. That is readily accounted for by the fact that sea serpents grow as well as anything else and a year should make considerable difference in one less than a century old.

Sandusky, Ohio, *Register*
July 8, 1898

Pennsylvania

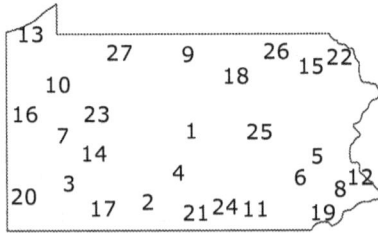

1. Bellefonte, Centre County
2. Bedford County
3. Greensburg, Derry township, Latrobe, and Monessen, Westmoreland County
4. Millcreek, Huntingdon County
5. Lehigh County
6. Windsor Castle, Berks County
7. Kittanning, Armstrong County
8. Montgomery County
9. Potter County
10. Oil City, Venango County
11. York County
12. Turkey Hill, Bucks County
13. Presque Isle, Erie County
14. Blairsville, Indiana County
15. Beaumont, Wyoming County
16. Slipperyrock township, Rose Point, and New Castle, Lawrence County
17. Somerset County
18. Utceter, Lycoming County
19. Clifton Heights, Delaware County
20. Riverview, Washington County
21. Mont Alto, Franklin County
22. Damascus, Wayne County
23. Punxsutawney, Jefferson County
24. Round Top, Gettysburg, Adams County
25. Shamokin, Northumberland County
26. Canton, Bradford County
27. Bradford, McKean County

A Curious Fact.

A large *black snake* was killed near this town which measured *eleven feet nine inches.*—It was first noticed by a slight crack which it made with its tail, not unlike the cracking of a horse-whip, and appeared to be in great agony—jumping up from the ground, twisting, coiling, &c. After it was killed, this was accounted for satisfactorily. Out of its mouth the tail of another snake was observed to be sticking: on pulling it out, it actually measured *five feet three inches.* This was the cause of the uneasiness in the living snake; having no doubt been partly strangled by its large mouthful. This great snake was long the terror of the cow-hunters in the neighborhood of the place where it was killed, and it no doubt would have continued so for a length of time, had it not been for its voraciousness which prevented it from running—It was fleeter than any horse; and bid defiance to the puny efforts of man to over take it.
Bellefonte Patriot.

Gettysburg, PA, *Republican Compiler*
August 16, 1820

Large Snakes.

Accounts of large snakes are going the rounds of the [papers] just now. The Pennsylvania *Argus*, Westmoreland county, this State, gives the following of one seen lately in that county:

Mr. Francis Cost, a resident of Ligonier borough, says the *Record* of that place, was down at Guffey's landing last week, and was told the following terrible account of a monster snake, then harboring in that neighborhood. He gleaned the following from Mr. Guffrey, who it is needless for us to say, is a man of undoubted veracity:

Near the "Yough" is a wild, rocky, rough hill, unfrequented except by animals, and is overgrown in many parts with bushes and briers. Some acquaintance of Mr. Guffrey, a respectable young man and a number of ladies went thither a few days ago, to gather black-berries. After arriving on the side of the hill, the young man separated from the ladies, and went higher up, where he soon found an abundance of ripe berries. Unconscious of the presence of danger, and rejoicing in his success, he proceeded for a while gathering the berries into a large tin bucket. All at once, about ten feet from where he stood, peering up above and among the bushes, he saw the head and neck of a huge snake, or rather serpent, fixing its awful eyes on him. Around its neck was a white ring, and its body was as thick as that of a man. For a moment, horror stricken, he was petrified to the spot—then involuntarily he raised his bucket, and with both hands dashed it at the head of the monster! Suddenly it darted at him, and suddenly he turned and ran for his life down the hill. Proceeding some distance, down the steep and rocky hill, he cast his eyes back, when he found to his dismay, that it was now close at his heels, and coming with great speed, its head erect, and its body thirty or thirty-five feet in length! Finding escape impossible in a direct line down the steep, as quick as he thought he sprang to one side, and ran in a different direction. Happily the serpent continued its course down the hill until it disappeared from his view.

He immediately repaired to the spot where he had left the ladies—with horror depicted on his countenance, and trembling in every joint. He related to them the horrible story of his adventure and miraculous escape from instant death. It is needless to say, the party immediately started for their homes, in a state of mind they never before experienced.

The fright so overcame the young man that he immediately went to his bed ill, and is yet confined, and under medical treatment.

One thousand dollars is said to be offered to the person or persons who will kill this hideous snake—$3,000 for it, if taken alive.

Since the above account was put in type, we had the pleasure of an interview with Mr. J. M. Miller, of Jacksonville, this county, who says that there is a monster snake in that region of country, and that its den is among shelving rocks. As evidence, he says, some time ago, a Mr. Taylor, a gentleman of veracity and courage, was driving his wagon in that rough region and passed an old field. Looking towards the fence, he saw the head, and about twelve feet of the body of an enormous snake, projecting through the fence. Its head was erect, and it seemed watching some object in another direction. He immediately stopped his wagon, and ran to the fence for a stake, with which to attempt to kill it. Unfortunately, as he took hold of the stake a noise was produced, which attracted its notice. Quickly it turned, and for a moment looked at him, and then turned and retreated through another opening into the field. Mr. Taylor testifies that he was within a few yards of his snakeship—and that if he had been so fortunate as to have had a gun, he could easily have killed it. He judged its length to be thirty feet, and its body at least six inches in thickness, or diameter.

Besides this, its track has been repeatedly seen as of a log dragged through grainfields, meadows, &c., in that section. Companies are organized, and they are now making a thorough search for the creature.

Huntingdon, Pennsylvania, *Globe*
Aug. 22, 1860

A Great Snake.

The vicinity of Rittersville, Lehigh county, Penn., has been thrown into a fever of excitement by the appearance there of a monster black snake, measuring from twenty-five to thirty feet in length, and the thickness of a common stove pipe. Last week she was come across by a lighting-rod peddler. His horse suddenly made a stop, and on looking about for the cause, he saw in front of him across the road the huge reptile, its head in a rye field, while its tail was just leaving the fence on the other side of the road. She followed her course through the rye, which was seen to sway backward and forward as she move through the field. The man hurried on and informed the neighbors who followed the snake with guns, but she took refuge between some rocks and was lost to view. Many other persons claim to have seen the snake at various times and places within the past year. Last year she was chased through a clover field, the path made by her course looking as if a heavy log had been dragged along. One of her favorite amusements was coiling her tail about a limb of a tree and swing to and fro like a large pendulum, darting her tongue in and out, snapping her jaws, and emitting a sound between a hiss and groan. In September last she was seen near Lehigh, slowly traversing a field, with head erect, and bearing in her mouth a large rooster which she had captured, and at another time a sportsman, of South Bethlehem, suddenly came upon her as she was in the act of catching a cat in her tightening grasp. The snake being of such immense size, and manifesting great rage, the young man became almost palsied with fear, and immediately ran away from the terrible locality, not thinking of his gun, which he might have used in destroying the monster. A party to go in pursuit of the monster has now been organized, when it is to be hoped the serpent will be killed.—*Allentown Democrat.*

Cedar Falls, Iowa, *Gazette*
August 25, 1871

A Battle with a Monster.

A young man named Franklin Rubright was on his way from his father's home to Windsor Castle, Windsor township, Pa. When about half a mile from the village, he met a monster black snake lying along the roadside. When within a few yards of the reptile, it made for him, and Mr. Rubright, having nothing with which to defend himself, was obliged to run. He soon found a club, however, and showed fight. After knocking down the snake several times as it rose up in front of him, the reptile seemed to be dead, and he endeavored to drag him, but the serpent soon recovered strength and fiercely attacked him a second time. The fight this time lasted several minutes, when the snake was again defeated, and fled to a neighboring corn field, where it was soon lost sight of, as it was about getting dark. According to Mr. Rubright's estimation the snake must be at least fifteen feet long and from four to five inches thick. Search was made for it the following morning, but without success. The snake has been seen at different times in the neighborhood within the last twenty or twenty-five years. Its hiding place, it is thought, has now been discovered, and a party of men are about to make a thorough search for it.

Smethport, Pennsylvania, *McKean County Miner*
September 10, 1874.

Big Snake.—Mr. Emanuel Bushman, of this place, communicates the following to the Baltimore *Sun*:

The now famous Round Top, a few miles from Gettysburg, years ago was the rendesvous of as large a snake as the Hall Springs snake of Baltimore city. I do not believe that Mr. Lee has exaggerated his large reptile. One sunny day in April of 1833 my brother, with six others, were exploring Round Top for the first time. The hill and surroundings were covered with a dense forest. As they were ascending the west side they suddenly came upon a monster snake, sunning itself upon the rocks. Part of them took to flight, but brother and two others stood to see how it would end. They described it as a black snake, apparently turning gray from age. Brother hissed the dog on it, and he thought from the capacious mouth that it would swallow the dog. They estimated its length to be from fifteen to twenty feet, and the thickness of an ordinary man's waist. They threw at it from above and it rolled down into its den. Father saw it before that about a mile from there at the big rocks, called the Devil's Den. Frank Armstrong saw it and was badly frightened. Grandfather saw it in his time, and mother says, the Indians used to speak of it as "heap big snake." Mr. Michael Fry, living near Round Top, saw it about thirty years ago, which is the last time I heard of it. How old do they get? Father says tradition traced it back one hundred years. Brother and Mr. Fry are the only two living who claim to have seen it.

Gettysburg, Pennsylvania, *Compiler*,
August 12, 1875

The Kittanning *Times* takes the cake for the biggest snake story of the season. Read and blush for the weakness of our erring neighbor: On Saturday last as the stage which plys between Kittanning and Dayton was approaching the Bealty farm, near the latter town, a passenger, Dr. Flower, suddenly called out "there's a snake."

The stage was at once stopped, and William Drake, the driver, jumped to the ground, and inquired where his snakeship was. The Doctor pointed to the side of the road, when Drake was astonished beyond measure at the sight which met his gaze. Two saplings stood about ten feet apart, and a huge black or blue racer, was stretched from one to the other, with his head and neck wrapped around one, and his tail entwined around the other. The Doctor was going to strike it with a stick, but Drake requested him not to, as he would try and shoot it. The first two shots missed, but the third one struck the reptile about the middle, when he dropped to the ground, and was soon dispatched. He was then attached to the rear of the stage and dragged into Dayton, where he was looked at by hundreds of excited people. He measured fourteen feet and one-half inches in length, and nine inches in circumference at the thickest part of his body. He certainly was a "whopper."

> Indiana, Pennsylvania, *Weekly Messenger*
> May 26, 1880

A Montgomery County Snake

There is a monster snake roaming around upper Salford township, Montgomery county, which is said to be fully thirteen feet in length by those who have peeped at it. The coat of the skin which this notable reptile shed last year was found to measure twelve feet six inches long.

> Williamsport, Pennsylvania, *Daily Gazette and Bulletin*
> August 3, 1880

The Boss Snake Story.—Mr. Solomon Smith, who is regularly employed at the Billmyer & Small Co.'s car shops, states that while visiting his parents, in Codorus township, several days ago, he accidentally found the hide of what must have been a huge black snake. The hide was stretched out on the ground, and accurately measured, with the following result: Length 14 feet and 3 inches, width 1 foot and 7 inches. Mr. Smith says that any one who may doubt this statement can go to his father's farm in old Codorus, and be convinced of its truthfulness.—York *Age*.

> Gettysburg, Pennsylvania, *Compiler*,
> July 18, 1883

Some Snake Stories.
A Monster Serpent Killed After a Long Fight.

Allentown, Penn., June 23.—A snake that measured 16 1/2 feet in length, and of proportionate circumference, and which has been pronounced a king snake by local naturalists, was killed by Elias Moser, a well known resident of Lynn Township, this county, a few days ago in the woods near his farmhouse. When Moser first saw the reptile only its head was visible between the rails of a fence. He supposed it was a large specimen of the ordinary black-snake, and, picking up a stone, threw it at the protruding head. The stone missed its mark, and the snake, instead of making off, began to make itself more prominent, and rapidly unfolded itself to the eyes of the astounded farmer. After a snake bigger than he had ever seen had come through the fence and the end of it was not yet, Farmer Moser turned and ran.

The snake followed him, and so closely that the farmer mounted a high stone wall and jumped down behind it. The snake glided up the wall also, and without delay came down on the other side. Moser saw there was nothing for it but to fight, and he grabbed a stone from the wall and hurled it at the serpent as it approached him. Fortunately the stone hit the snake near the head, which the reptile carried high above the ground, and knocked it down. Before the serpent recovered Moser seized another stone and tossed it square on its head, fastening it to the ground. The weight of the stone prevented the serpent from releasing itself, and Moser took advantage of the situation to hurl other heavy stones on the one that held the snake fast until the gradual ceasing of the twirling and coiling of the great body indicated that the reptile was overcome.

Moser did not venture to remove the stones, however, until he went home and got his hired man and a gun. When he returned thus reinforced the stones were taken off. The snake was dead. The head was long and flat. The upper part of the body was a bluish black, except two broad white bands around the neck. The belly was yellowish white. The great size of the snake is something unheard of for serpents known to abound in this latitude.

New York, New York, *Times* (excerpt)
June 24, 1887

On Friday of last week, Mr. Chas. Woelfling, a blacksmith of Galeton, Potter county, and a companion whose name we did not learn, were on their way home through Black Forest, from Young Woman's Creek, when suddenly a black snake, resembling a medium sized tree in bulk and length darted across their pathway, not three feet in front of them. Mr. Woelfling's companion with a yell of fright, took to his heels, but the blacksmith, grabbing up a club which happened near, fought the reptile and vanquished him after a "pretty tough struggle," as he expressed it. Hitching a piece of sapling to the monsters body, he "snaked" him to Galeton, where, upon measurement by Mr. Edgecomb, of the Ansley house, the reptile found to be fourteen feet and inches in length. The above is doubtless, the largest snake ever killed in the state. A den of the monsters is said to exist in that part of Potter county.— Port Allegany *Reporter*.

Olean, New York, *Democrat*
July 19, 1888

Snake Stories
Jeannette Reports More but a Greensburg Paper Tells the Biggest One.

Last year, avers the Greensburg *Record*, a monster snake was seen in the neighborhood of Donnell's mill. Its length was variously estimated at from 8 to 15 feet. An attempt was made to kill it, but it went into a thicket.

Nothing more was seen of this monster until on last Friday, when Miss Ida L. Robinson saw a cow down and struggling, and went to see what was the matter. Imagine her surprise to see, coiled around the neck of the cow, an immense snake some four inches in diameter and at least ten feet in length. The cow was being choked to death, and the brave girl took hold of the snake with both hands, and with great difficulty got it loose, or perhaps frightened it until it uncoiled itself and ran into the thicket. The snake was of a brown color and showed no signs of striking or biting, hence it is believed to be an anaconda, which has made its escape from some menagerie.

Miss Mary Hook, of Jeannette, and her guest, Miss Maggie Conley, of Allegheny, were walking along the road near Jeannette a few days ago when they came upon a large snake sunning itself in the middle of the highway. Each procured a club, and after a fight with the snake killed it. Miss Hook had a narrow escape, as the snake at one time during the fight had her within its folds. She was rescued by Miss Conley, who cut its tail off with her pen-knife. The snake was carried to Jeannette, where many people saw it. It was 10 feet 6 inches in length. First case on record in which women did not run from a snake.

Workmen excavating for new glass tanks in Jeannette a few days ago found a snake imbedded in the solid rock. It was 37 inches long, 5 inches in circumference, and weighed 11 pounds and 13 ounces. It is dark in color, and has three rings of a grayish tint immediately behind the head. It was dead when found and is still dead.

Connellsville, Pennsylvania, *Courier*
June 19, 1891

Here It Is.

The Oil City *Blizzard*'s annual snake story is this year as follows: For a number of years annual reports have come in from the vicinity of Fertig Station of a big black snake in that section. Its length has figured in such reports at from twelve to sixteen feet. One man based his calculation, he said, on the fact that he saw the big reptile sunning itself on a pile of twelve foot rails and that it extended about a foot beyond each end of the rail pile. It could have been killed on several occasions, it is stated, but for the reason that it was desired to capture it alive, and no one in the vicinity was equal to the task. It was seen a number of times last year but thus far this season has not made its appearance. Oil City's snake hunters have been notified that the first time it is seen this year word to that effect will be sent to Oil City, and it has been agreed that a delegation of expert snakists will respond to the notice and endeavor to capture the reptile.

Olean, New York, *Democrat*
June 2, 1892

Tricks of a Snake.
The Reptile Engages in the Dairy Business in Pennsylvania.

Al Palmer, of Derry township, brought to Blairsville, Pa., last Saturday the following remarkable story: He says that last Friday evening he was passing Mr. James Dunlap's barn at just about dusk when he noticed that there was quite a commotion among a herd of cows confined in the yard. One cow in particular seemed very much excited, and Mr. Palmer, thinking that the matter needed investigation, jumped over the fence, but immediately jumped back again when he saw an enormous snake hanging to the cow's udder.

He procured some stones which he threw at the snake and succeeded in making it dislodge its hold upon the cow, when it started off toward Deer hollow, which is not far away. Mr. Palmer says that the snake was fully 14 feet long and that its diameter at the thickest portion of its length could not have been less than 6 inches. He also claims that it left a luminous wake or trail behind it, and that it was probably a black snake or one of a similar species. An investigation of the herd of cows showed that the huge reptile had robbed five of them of their milk. An organized search for the monster will probably be made, and it is hoped that it may result in the serpent being put out of the way.

Hagerstown, Maryland, *Herald and Torch Light*
May 3, 1894

Thought It was a Bike.
Pennsylvania Bark Peelers Find a Queer Trail.
They Concluded It was Made by a Bicycle, but Later Developments Showed Them that It was the Work of a Snake.

James R. Zepp and James C. Meyers, of York county, Pa., are doing a job of bark peeling in that county near the Maryland line. Sunday afternoon they were walking along the Melrose road on their way to the woods when they came upon a track in the road.

"Who do you s'pose can be going through this country on a bicycle?" said Zepp.

"Somebody that must have muscle and nerve," replied Meyers. "He must have a tire on his wheel like a lumber wagon, from the width of the track it makes." They measured the track. It was six inches wide. It followed the road for two miles and then turned off toward the woods and ended. Then for the first time it struck the two bark peelers that the track had begun in the road just as suddenly as it came to an end. They thought over the puzzle for some time and could come to but one conclusion.

"It's a spook bicycle," said Zepp, and Meyers agreed with him.

They went on with hurried steps and had gone a mile or two, when suddenly an enormous black snake glided out of the woods into the road. It had a cock pheasant in its mouth and was dragging it along by the neck. The snake kept the road and traveled so fast that it was out of sight before Zepp and Meyers had recovered from their astonishment. Then they saw that it was not a bicycle that had made the mysterious and spooky track in the road. The wide track was made by the enormous black snake.

It was some time before the two bark peelers could make up their minds whether to go on or to return to Melrose. They finally armed themselves with heavy clubs and went on. The great trail followed the road for nearly a mile, when it turned toward the woods again and disappeared. Zepp find Meyers were now within two miles of their camp where they had a gun, and they hurried on. Within half a mile of their camp the big snake came into the road again like a flash. This time it had a rabbit in its mouth.

"Lord!" exclaimed Zepp. "He's gulped that big pheasant and han't had enough yit!"

They gave the snake time to get a long distance in the woods, and then they broke for the camp on the double quick. They found Andy Flite, of Cumberland, there. Andy is a fox, coon and possum hunter, and when he heard about the big snake he called his dog and started out to see if he could not make a bit of a snake hunter, too. Zepp and Meyers took their guns and went along.

They took the trail where the snake crossed the road, and Andy's dog followed it as if it had been a fox scent. He led the men a mile into the woods and came to a noisy stand. He barked and growled and yelped so vociferously that Andy exclaimed:

"That's the oncommonest snake, I reckon, that was ever treed round hyar or you wouldn't find old Dan cuttin' up like that. He's holed up a dozen big rattles to wunst 'fore now and never made no setch fuss."

The hunters drew near to the dog and immediately saw that he had good cause to make a fuss. Two snakes lay stretched out on the rocks. They were black as coal or they could easily have been mistaken for chestnut saplings lying on the ground, so Zepp and Meyers declare. Each snake seemed swollen to an enormous size just below the neck. Andy Flite shot both of them through the head, and they thrashed around so in the bushes that the dog and the two bark peelers ran away.

The snakes finally gave up and Andy cut them open. In one was a cock pheasant and in the other a rabbit. The snake that had swallowed the rabbit also had a mule's shoe in its stomach. This led to the report when the news of the killing of the two snakes was carried to Melrose that one of the snakes had swallowed a mule. This was found out later to be untrue.

George Sampson identified the shoe as one he had tied around the neck of a cat he had thrown into the mill pond a couple of weeks ago to drown. The cat had evidently been too much for the weight of the shoe, had escaped from the pond and run up against the snake, which at once took it in, shoe and all.

The smaller of the two snakes measured 14 feet, the other 14 feet 9 1/2 inches. They were larger by four or five feet than black-snakes usually found in that neighborhood.

       Monroe, Wisconsin, *Evening Times*
       August 11, 1894

One of the most delightful places to visit on bright autumn days is Turkey Hill. Its woods, filled with the most beautiful tinted autumn leaves and rare plants, delights the student in botany, and the study of the peculiar formation of the hill makes it equally as fascinating for those who are interested in geology. Or seated on one of the grassy terraces of its eastern slopes lulled by the gentle rustle of the leaves in the adjoining woods, and the music of the birds, one might sit for hours and gaze upon the beautiful panorama, the fertile and well-cultivated Penn's Manor farms, enveloped in the mists of the Indian summer day, the course of the Delaware marked by a streak of smoke, probably from one of

Captain Edwards' river steamers as it rounds Periwig and heads toward Trenton, the steeples of Bordentown in the hazy distance, the Jersey pines, and beyond, the place where the sky seems to come down and touch that highly favored little State. This is indeed a picture to inspire a poet, but none will probably be inspired for some time to come; neither are any of the treasures of the vales and woods of Turkey Hill likely to find their way into any collector's cabinet, all owing to the fact that an innocent snake is now roaming at will over that beautiful spot. This snake is said to be "nearly 20 feet long and nearly as big around as a telegraph pole." It was seen about a week ago by the driver of a bakery wagon from Morrisville. It then disappeared and was not seen until Sunday when Archie Wheelock, an Indian boy, was going up the road that leads over the hill and through the thick woods. Suddenly he heard a strange but beautiful whistling noise and turning around the sight that met his eyes he will not be likely soon to forget, for there right behind him in the road was the huge monster, its head elevated about five feet from the ground, its eyes flashing fire and its forked tongue playing like lightning around the corners of its mouth. Archie is a brave boy and came from the far West, where ferocious beasts abound, but he never saw anything before that scared him like this snake, and he turned and ran and never stopped until he reached the home of his employer, Mr. B. F. Muschert, more than a mile away. Mr. John Carter, who resides on Turkey Hill, says this huge snake was first seen on the hill about 80 years ago, and has been seen about every 5 or 6 years since that time. Where it stays during the time between its periodical visits is a mystery. It is said a number of smaller snakes measuring from 8 to 10 feet in length have recently been killed in Patterson's meadow adjoining the base of the hill.

Bristol, Penn., *Bucks County Gazette*
October 27, 1898

As Big as a Stove Pipe.
And Fourteen Feet Long was the Size of a Snake Seen on Wolf's Hill. (?)

[Written for the *Compiler*.]

Some years ago I heard a snake story told by a citizen of our town, which was quite an experience and he declares kept him back in his growth one year.

With several friends they decided to take a hunt on Wolf's Hill. When near the thick pines they flushed a pheasant which flew into the pines. They then decided that one should go on each side of the thicket and one through the thick pines and brush and make all the noises possible so as to chase the bird out. The [task] fell to our friend (who is seen on the streets almost any time in the day) to go through the brush. And here is his story as told in the cigar store that evening.

"I could not carry my gun so I left it with one of my friends, I started through the thicket. After going some distance I found the only way to get through was to get down on my hands and knees. I had not proceeded in this way very far before I saw not over 2 ft. from my nose an immense snake thirteen or fourteen feet long and about the size of a stove pipe through. My hat hung suspended by my hair above my head a foot. I started to back out, and had not backed far when I ran against a tree, I thought there were snakes all around me, how I ever did get out I cannot tell. I thought the snake had me for sure. Scared! Well, I guess. I would not have gone back into that thicket for a thousand dollars. If that snake is still living out there, he must be as large as one of those telephone poles." L. F.

> Gettysburg, Pennsylvania, *Compiler*,
> Unknown date.

From *Science* (Moseley 1901):

The Python in Pennsylvania.

To the Editor of *Science*: On August 9, a python, probably *Python natalensis*, was found in the grass on Presque Isle, Pa., by three young men from Erie who, as they supposed, killed it and took it to the city. However, it revived and was exhibited in the window of the Tribune bicycle store. On August 29 I measured and weighed it. The length was about seven feet four inches, greatest girth eleven and one-half inches: weight, seventeen pounds. That evening it pushed away the wire to netting from one corner of its cage, and escaped. It probably took up its residence under a building in the rear of the store, but had not been seen when last I heard, October 14. Reports of the liberation of large snakes in the vicinity of Presque Isle I investigated, but they proved to be unfounded. Who can tell how this African snake found its way to the shore of Lake Erie and how long it had found subsistence there?

E. L. Moseley.
Sandusky, Ohio
Oct. 27, 1901.

Terrible Serpent at Blairsville.

There is said to be terrible consternation among the employes of the Columbia Plate Glass works at Blairsville, owing to a monstrous serpent which exists in a swamp near the factory. According to report it has horns two feet long, is six inches in diameter and fifteen feet long, and hisses like a locomotive blowing off. The serpent is said to be shaped like a pint bottle and the employes are facinated [sic] by the peculiar hiss and can hardly be kept away from the spot. No efforts to kill it have been made lately.

Indiana, PA, *Weekly Messenger*
August 27, 1902

—Mrs. Lewis Dershimer, while driving the stage m place of her husband, from Beaumont to Tunkhannock, killed the largest snake seen in the upper part of Pennsylvania in many years. The horses were trotting along the road in Eaton township, when they shied at the snake which was coiled and had raised its head four feet above the road. It was a huge blacksnake and was full of fight. Dershimer always carried a loaded shotgun on the stage and Mrs. Dershimer, instead of being frightened, reached for the gun, took careful aim and killed the snake. It measured just one inch more than 11 feet.

Wellsboro, Pennsylvania, *Agitator*
September 10, 1902

A Millcreek Monster
Snake Larger than Width of Wagon Road in Mercer

Millcreek, Pa., July 8.—That Millcreek blacksnake is no fake, but a "veritable fact," and J. F. Davis offers a bounty of $10 for him, dead or alive, and the amount will be doubled by other citizens. The "varmint" has been seen by several persons of truth and veracity at different times for more than 35 years, and its skin has been frequently found, but the "real thing" has so far escaped. The last one to see it was Jacob Crouser, a man of truth—superintendent of the Presbyterian Sunday school at New Lebanon—and not afraid of ghosts. He declares that the reptile was lying across the highway with both ends lost to view in the rank weeds and brush. Whether he walked or ran, or what he did, deponent sayeth not.

Oil City, Pennsylvania, *Derrick*
July 9, 1906

Caught Eagle and Big Snake
Slipperyrock Township Man had Two
Exciting Combats in One Day.

A great bald eagle, the first of its species seen in this locality in many years, and a huge blacksnake measuring 17 feet in length, were captured one afternoon by M. C. Gallagher of Slipperyrock township, after a desperate fight. His struggle with the eagle occurred first, and the man did not come unscathed from it.

Gallagher sighted the great bird that stands as the emblem of its country, and drew a bead upon it, with a shotgun. The eagle dropped at the report and lay as if dead. Supposing it had been killed Gallagher rushed up to seize its prize.

The eagle came to life when it was picked up, evidently having only been stunned by the fine shot in Gallagher's gun. Then ensued a desperate fight between the man and the great bird, which fought with talons and beak to avoid capture.

Gallagher's clothing was badly cut and torn and his face and hands were scratched before he was finally able to stun the eagle with a blow from a club. That was an experience that

seldom occurs to any man, but Gallagher had another immediately after.

While he was tying the captive bird, a small boy came running along the road and very excitedly told him there was a
huge snake a little further down. Gallagher went back with
the lad and found a large blacksnake, lying stretched out in
the dust. Seizing a shovel the boy carried Gallagher attacked
the serpent, which wrapped its coils about the shovel handle
and was then captured. Both eagle and snake have been taken
to the Pittsburg zoo, where they are living in captivity.

> New Castle, Pennsylvania, *News*
> August 29, 1906

A monster blacksnake, 12 feet long and 33 1/2 inches in
circumference, was shot by citizens of Somerset. They saw him
emerge from a great hole beside one of the graves at the rear
of the little Brothers Valley church. It is the opinion that his
snakeship is one of the South American and Indian serpents
that escaped from a circus several years ago.

> New Oxford, Pennsylvania, *Item*
> November 16, 1906

Who Will be the Next to See Rose Point's Monster Snake

Rose Point is famous for its blackberry patches, its groundhogs and its general air of prosperity, but now a new fame has
spread throughout the countryside for one of its sons has become
famous.

John Johnson is his name, and though his neighbors have
seen him daily in his travels about the large farm of the late Judge
James P. A..., where he struggles with the large crops and the
daily prognisticators at the weather bureau, not one of them ever
suspicioned him of having any of the qualifications that go in
making a man of fame.

As is usual in such cases, the fame was thrust upon him.
Rather, according to his own version, it pursued him, and it

was only a very pretty exhibition of sprinting that it failed to completely envelop him in its coils.

And this is how it happened. Last Friday Johnson started out after a piece of timber at the [far] end of the old cow pasture to get some hard wood for the winter's burning. Incidentally he wanted to look at the chestnut crop, which from all ... would be almost a total failure.

He worked for an hour wrestling with the hickory and then started for the chestnut crop. Passing through a clump of brush that fringed a [hill] upon which several of the trees stood. Johnson saw a movement in the [bushes] which looked to have been the passing of a rabbit. He made a kick at the bush to startle the animal. It was a successful kick, for before he ... around to climb the hill the huge head of a monster snake rose directly in front of him and swayed for a moment over the tops of the scrub.

Johnson stood looking into the ... eyes for a brief moment, than, as the snake made a [sliding] motion with its head, the meaning of which .... suddenly through his mind and he ... down the hill. Out across the meadow he sped with never a look behind, for the ... swish of a huge body sawing its way through the grass told his strained ears that snakey was still in the race.

Johnson made home in record time. Where he left the snake he does not know, but of this — he is certain. That the snake was 25 feet long if an inch. It was black, too, and its head was as large as the three-pound prize .... that Jim Bunkle brought to the Mercer fair this fall.

And now folks, even those who have known Johnson a life time, are saying things about him. But he doesn't care, for he saw it, felt the hypnotic glance of its eye, heard the swish of its body in the grass, and besides, if anyone wants to they can go up into the ... pasture near the line fence where the soft earth still holds imprints of the flying boots and, if one looks closely they might still see many ... places in the field where the monster ... glided along.

That is the substance of the tale as told by one of Johnson's close friends and now they are saying he is the most famous of all their famous men, though they didn't say men.

New Castle, Pennsylvania, *News*
October 23, 1907 (Some portions illegible.)

While gathering huckleberries on the ridge above Young-stown, near Latrobe, Pa., Edward L. Bates was attacked by a monster blacksnake. It dropped upon the boy's shoulders from the limb of a tree and began wrapping its coils around his body. Two companions succeeded in killing it. The snake was 9 feet 10 inches in length.

Indiana, Pennsylvania, *Progress*
August 19, 1908

Monster Snake Killed

A battle royal was waged on Boher's Hill Monday after-noon when the progress of George A. Calhoun, who was leisurely strolling over the wooded knoll, communing with nature in the deep shade, was impeded by a monster blacksnake. The venomous reptile drew itself into an hyperbolic curve and immediately showed fight. Mr. Calhoun took in the situation. Seeing that retreat was impossible he gathered together all possible primitive munitions of war, and, surrounding his antagonist, began to hurl the rugged missiles, finally lower-ing the proud head of the writhing, coiling snake. A post mortem examination showed numerous spots on the ebon coat, thus indicating that Mr. Calhoun's every throw was effective. A measurement revealed the fact that the snake was eleven feet two inches long. The measurement was checked and verified by Pardy Gilchrist.

Bedford, Pennsylvania, *Gazette*
May 28, 1909

A Monster Snake
Fourteen Foot Blacksnake Killed at Utceter Saturday.

A blacksnake measuring 14 feet and seven inches in length was killed by a freight train at Utceter, two miles below Slate Run, on Saturday. The monster reptile was cut into three pieces. Daniel Callahan, on whose farm the reptile had been seen several times of late, gathered up the carcass of the snake, placed the pieces together and measured its length, the size aggregating the remarkable figure given above, says the Williamsport *Sun*.

Some contend that the snake is not a native species ... [missing] ... was seen gliding through a hay field, with its head raised a foot or more above the grass. The section men on the New York Central railroad became so alarmed over the report about the big snake that they feared to work in the locality of Utceter.

The snake had gotten partially across the railroad track Saturday when caught by the wheels of a New York Central freight engine. The snake was taken to Cammal and put on exhibition.

To add to the excitement over the huge snake, it is reported that one fully as large has been seen on the James B. Tome farm, across Pine creek from Slate Run.

Perhaps one of these big snakes chased Mr. Dennis W. Navle, of Wellsboro, recently, an account of the thrilling occurrence appearing in the *Agitator* of June 16.

Wellsboro, Pennsylvania, *Agitator*
June 30, 1909

Big Snake at Whiskey Run
Monster Reptile Shows Discrimination in Choosing Abode.

Clifton Heights, PA, April 15—Displaying fine serpentine discrimination by making his habitat at Whiskey Run, an enormous black snake is spreading terror through the farming district of Springfield Township, Delaware County. The snake is described as at least 12 feet long with a peculiarly large head and a body of great girth.

Two of the latest observers of the reptile are Robert Lacy and George Sullivan who met it in the road. Yielding the right of way, the snake retreated to a field and crawling into a hollow tree, pushed its held out of a hole some distance up the trunk and hissed, making a most peculiar noise. The men made no effort to kill the reptile, as they had no weapons, but ran and told their neighbors of what they had seen. Old farmers m the township say they have known of a monster snake in the township for years and believe this is the same one. Close watch is now being kept on the tree and its surroundings in an effort to capture the snake.

Trenton, New Jersey, *Evening Times*
April 15, 1910

This Snake Was Nine Feet Long

Beaver Falls, May 18.—John Corcoran, ex-road master of the Pittsburg and Lake Erie railroad, killed a snake on his farm in Riverview the other day, which he claims is the largest snake ever killed in the county. The snake was a blacksnake and measured 9 feet and 4 inches from head to tail. Mr. Corcoran killed the snake with a fence rail after a terrific battle.

New Castle, Pennsylvania, *News*
May 24, 1911

Monster Snake Seen

While George B. Shaffer, who resides near Mont Alto and supplies milk to the residents of that town, was out in the field looking after his cattle he saw a black snake between 12 and 15 feet long.

Mr. Shaffer estimates its length by a lane across which it was stretched. It reached from one side to the other and was as thick as an ordinary fire hose. He didn't carry his investigation very far, fearing that his snakeship might resent too close inspection.

This snake has been reported before. It was seen about a year ago and is believed to be the same snake for which John Robison's circus men hunted for a week about 30 years ago.

Gettysburg, Pennsylvania, *Adams Co. News*
May 27, 1911

Harry's Big Snake Story

Harry Guist is a brave man, a modem hero, a martyr. He came to town-last night with the biggest snake story of the season. While returning home from his day's labor on the Shepler farm below town he encountered a monster black reptile.

With a club as big in size as the snake Harry started in to dispatch the serpent. The more he tried to kill him the harder the snake tried to live and it was nearly an hour before Harry came out the victor. It measured just ten feet and ten inches in length when stretched across the road, says Harry.

> Monessen, Pennsylvania, *Daily Independent*
> July 18, 1912

Some Snake This—Thirty Feet Long?

Damascus, Pa., Aug. 11.—Probably one of the greatest sensations that has visited Damascus in many a day was launched one day two weeks ago when Fred Greeley, a reliable and truthful man, who lives between Tyler Hill and Galilee, came to Tyler Hill and told how he had seen snake that was between twenty and thirty feet long and was as big around as an ordinary stove-pipe.

Mr. Greeley said that he was out picking blackberries on his farm, which is known as the old Greeley place and is situated somewhat southwest of the main line between Tyler Hill and Galilee, and while walking around among the bushes he started to part two bushes to go through them, and in doing so he saw a log lying on the ground and was just in the act of stepping on it when the apparent log commenced to move. He ran back aways and then saw that it was a monster snake, or Boa Constrictor, and that it moved along lazily with no attempt to harm him nor to run away. It gradually made its way into the thick undergrowth and disappeared, and the trail it left was just the same as though a large log had been dragged along the ground.

Mr. Greeley's story was called a joke by everyone who heard it except some of the old timers around Tyler Hill, who have heard of this monster before and took considerable stock in the story. Several years ago Sidney Brush, and his sister were coming from

Milanvllle along the Frosty Hollow road and they told at that time of seeing an extremely large snake along the road, but from that time to this it has never been seen, or at least it has never been reported. Possibly it has been seen, but no one had the nerve to say that they had seen it for fear that they would be ridiculed.

It seems a pretty big story to swallow, but facts seem to bear out Mr. Greeley that such a snake exists in the wilds of northern Wayne. About fifteen years ago a small circus that was traveling over the Erie had a small wreck between Cochecton and Narrowsburgh, and the car that contained the snakes was partly smashed up and a. number of the snakes escaped, among them three large ones known as Boa Constrictors. They went into the woods and for days efforts were made by the circus people to find the snakes, but there were only a few of the smaller ones caught and the big ones remained in the woods. The probability is that one of these monsters is the one that has been seen in Damascus a number of times.

Middletown, New York, *Daily Times-Press*
August 13, 1913

Black Snake More Than Nine Feet Long Is Shot
Monster Reptile Puts Up Fight for Life Until Killed by Bullet.

Punxsutawney, Pa., July 24.—A monster black snake was killed on the Hugh Neal farm, near Punxsutawney, Jefferson county, by Roy and E. E. Lettle. It measured nine feet two inches in length and eight inches at its thickest circumference.

The men were passing through a little timber tract when they discovered the monster reptile coiled in the remains of a tree that had been blown down. They closed the end of the hollow log and made an aperture for the snake to crawl out of while they waited with a revolver. The snake, after being aggravated, made a dive for Roy Lettle's leg and got back safe into the log. At the next dive the reptile was shot through the head and killed. The skin of the snake is on exhibition here. It is attracting a great deal of attention.

Syracuse, New York, *Herald*
July 25, 1915

Another Round Top Snake Story;
This Time Reptile 14 Feet Long
"Sol" Pittenturf Runs onto "Whopper" in
Locality Noted for Big Ones

Clear the slate for the prize snake story of the year. 'Tis true the season for snakes has just commenced, but it is doubtful whether a reptile of larger proportions will sneak from his winter hiding place in this locality. The scene for this snake story is Round Top, where "whoppers" have been reported for years.

"Sol" Pittenturf, whose truthfulness we do not doubt, though he is an ardent fisherman, and who does not imbibe freely in the liquid that gives rise to some snake stories, is authority for the statement that a blacksnake 14 feet long inhabits the rocky east side of the hill. Frequently in past years the story of a huge blacksnake was spread—usually about once a season.

John Rosensteel killed a reptile 11 feet in length some time back, and it was thought the source of all the tales about "big ones" was in snake heaven. Then here comes Mr. Pittenturf with another of zoological garden proportions. He was within a few feet of the snake and for that reason insists that the size he gives is accurate. He made an effort to kill it, but the serpent glided into the bushes and among the rocks, and further efforts to rout it from its hiding was fruitless.

For many years Round Top, in the wildest sections of the hill, has been noted for big blacksnakes and not infrequently one of the big ones is stirred out.

Gettysburg, Penn., *Star and Sentinel*
April 28, 1917

Fights Big Snake For Life
Twelve-Foot Monster Attacks Man and Battle Lasts 20 Minutes.

Greensburg, Pa., Aug. 4.—A big black snake, measuring within a fraction of twelve feet in length, gave G. K. Johnston, of Mill Bank, the fight of his life.

Johnston was near Bell's station on the Ligonier railroad when he found it necessary to cross a rocky gorge adjacent in the Loyal Hanna Creek to take a short cut for his home. The big snake, coming from the underbrush, entered the gorge at the same spot and time as Johnston. And then the battle began.

With a stout cudgel some five feet long, the man essayed to beat the reptile off, but the latter refused to give up the fight. Several times the snake narrowly missed getting his coils around Johnston.

They battled back and forth in the ravine for fully twenty minutes, and exhaustion was overcoming the man when the snake succeeded in getting his head within a few inches of Johnston's body. Striking downward with all his might, Johnston succeeded in dealing the reptile a blow on the head which stunned it. With a few well-placed blows the snake was then killed. Its body was the thickness of a man's arm.

Frederick, Maryland, *News*
August 4, 1919

Foresters Relate New Snake Story; May Be Pet Python

Harrisburg, Sept. 4.—The South Mountains of Pennsylvania are notable for two annual productions. The one is a bumper huckleberry crop, the other an almost, but never quite, improbable snake story. Of the latter, the newest to be reported concerns an enormous snake, whose length has been reported by various observers as from seven to twelve feet, which employes on one of the roads now being built on the Mont Alto State Forest claim to have seen.

The snake is believed to be a python, and is said to have escaped from its owner, who kept it as a pet at his summer home in the. mountains near Pen-Mar, Md. Forest rangers report that rarely are native snakes more than five feet long encountered in this region. These are usually the non-poisonous pine snakes or black snakes, though an occasional large rattlesnake is found. The other poisonous species native to Pennsylvania, the copperhead, rarely exceeds three feet in length.

If not found and killed before cold weather arrives, the python is not expected to survive the winter since it is a native of the tropics and would succumb to the rigors of freezing temperatures.

Huntingdon, Penn., *Daily News*,
September 4, 1931.

Roaming With Richards

It isn't really summer, until somebody turns in a snake story, and it isn't winter until they quit turning them in. Clint Carroll gets credit with opening this summer with [a] snake story that isn't half bad.

He was working out among his evergreens on the front lawn and he has some beauties by the way, when a snake's head nearly the size of a teacup pushed itself out of the shrubbery. Now snake heads that size in this part of the country are rare, and much rarer if you're sober as Carroll was.

Just a flash of the head was all he had but that was enough. Mental pictures of what might happen flashed through his mind. A snake that size would consume a dog anyway. It might be a boa constrictor or a python escaped from a circus, it might be that Rose avenue was still infested with snakes. It called for some help.

One of the neighbors came over and between them they beat the shrubbery for a while, and out comes the snake with the big head. It was a garter snake that had picked off a juicy toad for a spring lunch and had not yet swallowed the toad. The feet were sticking out of its mouth, and the snake's head was distended to twice its size.

And so the spring snake story season is open and Carroll, a post office employee, gets credit for opening it. And the only reason we're running it is that we know he is a sober sort of citizen. Otherwise...

New Castle, Pennsylvania, *News*
May 23, 1933

Monster Snake is Reported Captured

(International News Service)
Shamokin. Pa., April 6.—Lee Erdman created considerable excitement here today by exhibiting a monster 11-foot snake he said he caught in a burlap trap on a refuse pile three miles north of here.

The brightly-hued reptile, about 5 or 6 inches in diameter, weighed almost 65 pounds.

Erdman said he believed it was a cobra which had escaped from a circus or carnival.

He set his trap after several boys reported seeing the snake on the refuse pile.

New Castle, Pennsylvania, *News*
April 6, 1935

Girl Finds Strange Snake Not Native of State

Canton, PA., July 2—(AP)—Marjorie Packard found an odd-looking snake, five feet long, tied a piece of twine around it and hauled it home. The snake is not native of Pennsylvania and neighbors said it evidently escaped from a circus.

Marjorie's father is wondering what to do with it.

Clearfield, Pennsylvania, *Progress*
July 3, 1935

Strange Snake Killed on Nearby Highway

Jack Hinckley of Bradford, who operates the Bradford-Smethport mail truck, created quite a controversy one day last week when his truck killed a large snake on the Ormsby-Mt. Alton highway, near the B. & O. underpass at Backus.

The snake is described as about three feet long, dark colored, with scales like a rattlesnake and a flat head.

One man who saw the reptile expressed the opinion that it was some kind of water snake, although there was no explanation how a critter of this species ended up in the highest and driest spot in the area.

Postmaster C. J. Parsons of Ormsby recalled that a rattlesnake had been killed in the hilltop community about 40 years ago and none had been reported since that time. It was the general opinion then that this reptile had been "imported" in some unknown manner.

Smethport, Pennsylvania, *McKean County Democrat*
August 2, 1951

The Broad Top Snake

The Broad Top serpent stories deserve recognition on their own. In a state where, historically at least, big snakes are pretty common in the folkloric landscape, the stories from Bedford and Huntingdon County manage to stand out. Besides producing more stories than any other part of the state (at least that have been published), the stories continue right up to the present day. I've mentioned my suspicion that an overlooked population of *Pituophis* is responsible, but of course that will require a specimen to confirm. What follows is a rundown of the sightings coming out of this region. Much of this comes from the writings of Broad Top *Bulletin* publisher/editor Jon Baughman (Baughman and Morgan 1977; Baughman 1987), who continues to record sightings as they occur.

1859—

A Formidable Snake.—Two Men Whipped.—The truth of the following snake story is vouched for both by the Cumberland *Telegraph* and Bedford *Inquirer*:

A few days ago, as the mail hack from Cumberland to Bedford was passing along about one mile south of the Half Way House, a large snake of a dirty black color was lying across the road. The driver, Mr. Samuel Bagley, drove the hack swiftly, both wheels running over it, but without apparently injuring the snake.—The driver and Mr. A. B. Cramer, of this place, then got out and fought it with a rail; it ran in a fence corner, and raised its head on the fence, the other part of its body being coiled up. Mr. Bagley got on the fence and struck it on the body with a stone of 25 or 30 pounds weight, which bounced off, apparently not injuring it. They not caring to risk themselves in any more danger, let the snake slide across the fields at the rate of 2:40. The snake was as thick as a man's leg, and from 10 to 15 feet in length.

Gettysburg, Pennsylvania, *Compiler*
July 11, 1859

1882—
Bedford County Had Its Snake Story, Too

(From time immemorial tales of monsters have made fascinating reading. Bedford county has its own snake story, the account of a huge serpent that was the terror of the Cumberland Valley region in which it was said to reside. Forthwith is an account of this fabulous reptile first published in the *Gazette* August 11, 1882)

Aug. 11, 1882
About 50 years ago, tradition has; it, a hunter who was plodding his weary way o'er the mountains of Cumberland Valley township, was suddenly startled by the scattering of dry leaves and the crashing of brush on the mountainside.

He ran quickly, thinking that he had startled up a deer, and upon reaching a point which commanded an extensive view, he beheld a monster reptile making its way down the mountain. It traveled with head erect and at great speed and seemed to be from 20 to 30 feet in length. The hunter did not care to linger long in that vicinity and he accordingly hastened home in double-quick time.

For many years nothing was seen of the reptile, but about 25 years ago, when stage coaches were still running between Bedford and Cumberland, Samuel Bagley, the veteran stage driver of this place, while driving through the valley one day ran over an obstruction in the road that almost upset the coach. He at once reined in his horses and upon looking back discovered a snake stretched partly across the road and with its head concealed in the bushes at the fence.

Bagley at first felt a little timid about tackling a snake of that size, but he finally summoned up courage, armed himself with a fence rail and advanced to the attack. He poked his rail very vigorously into the bushes and when he felt the reptile's head had been pounded into a "jelly", he preceded to make an investigation. Imagine his terror when he discovered that the smaller part of the snake was in the roadway and that the head, instead of being concealed in the bushes, was down about the middle of the meadow, elevated at least six feet from the ground, with flashing eyes and a spear-like tongue that constantly darted in and out of its mouth. Bagley can't remember how he got back to his coach, but he does remember that for the next five miles his horses traveled faster than they had ever traveled before.

The snake was again seen five or six years ago by John Appel of Cumberland. He described it as a blacksnake, between 25 and 30 feet in length and 9 inches in diameter at the thickest part. It was next seen by Perry Morgan of Colerain township who found it sunning itself in the road one day. He was obliged to leave the public highway and drive through the woods in order to get around it. His description tallies with that of Appel.

The snake has always been seen in the vicinity of Johnathan E. Luman's place. It comes from Evitt's mountain, above Luman's house, crosses the road leading from the Cumberland Valley road to Zembower's mill and makes its way to Knobley's ridge where it goes, it is thought, for the purpose of obtaining water. On the 28th of July it was seen by John Hardy, a peddler from Juniata township. He described it as being about 25 feet long and as thick as a man's leg at the thigh. On the same day L. P. Bruner saw the track of the serpent in the roadway and the tracks of four smaller snakes that accompanied the large one. On the following Monday, James Zembower, Luman and Bruner armed themselves with shotguns and started in search of the monster, but they hunted diligently all day, they were unable to capture the prize. They succeeded, however, in killing one of the smaller snakes.

This dose of snake story is intended for warm weather reading. It should be imbibed with moderation and generously, seasoned with snake medicine. Otherwise the nerves might not be able to stand the strain.

Bedford, Pennsylvania, *Gazette*
June 3, 1949

1907—

30 Years Ago

A monster snake which has been said to have been seen several times in the past three years in the vicinity of Ray's Cove, Bedford county, was seen a few days ago by two young men, Howard Rhodes and Francis Foor, while plowing on the Judge Foor farm. According to all accounts the snake must be fully fifteen feet in length and at least six inches in diameter.

Huntingdon, PA, *Daily News*
July 23, 1937

1927—

A witness stated that in August of 1927, "I went to town for groceries and was returning home. I sat down to rest near the old mine at Putt's Hollow, that's on the old road from Hickory Hill to Saxton. I heard a noise and thought several dogs were badgering a woodchuck. Then things got quiet and I heard a noise coming through the leaves towards me. I waited and watched; suddenly it appeared. About 20 yards away, a giant snake appeared and crossed the road. It was so large that its head was on one side of the road and its tail was on the other. It was gray in color with yellow markings around its face. I followed it down the mountain and it dislodged rocks and made a definite track in the leaves. As it neared the water, I climbed back up the mountain." (Harrisburg, Pennsylvania, *Sunday Patriot-News*, October 23, 1977)

1933—

A giant snake was seen on Saxton Mountain (Baughman 1987). In the same period, three fishermen on Roaring Run, between Saxton and Riddlesburg, encountered a huge snake. They didn't stick around to investigate (Baughman 1987).

pre-1940—

Jon Baughman (1987) received a letter from a former resident of Six Mile Run who saw a large serpent while visiting relatives in Kenrock, Carbon Township, Huntingdon County. The man said that while fetching water, he had to cross two sets of railroad tracks. "On one occasion approximately half way, there was a snake stretched completely across both tracks plus the right of way between them. My recollection is that it was at least 20 feet. I can recollect the snake was 8 inches to 10 inches in diameter, and moving slowly. My sister screamed. I don't remember if I did or not. After what seemed like a long time, the snake passed, and we continued on to the spring for water."

1950s—
A man was driving between Robertsdale and Cooks near dusk, when he ran over what he thought was a large log. When he stopped to check, the "log" crawled off into the woods (Baughman 1987).

Two Coalmont men drove around a large "log" near Hickory Hill. When they stopped to investigate, they realized it was a large snake. They estimated it at 25 feet, and they couldn't see the head (Baughman 1987).

1957—
A large snake was seen by a couple of mushroom hunters on an old farm above Riddlesburg. They could see about six feet of the snake laying across a rock, several feet more of it on both sides, but couldn't see the head or tail (Baughman 1987).

A Coaldale man, walking his dog, encountered a large snake in a briar patch, frightening the dog badly (Baughman 1987).

1957—
Two miners in the Hickory Hill area were moving equipment when they found a snake whose body was as large as an 825 tire. One man hit it with an axe, and it took off, ending up in a large log. The next day, one of the men returned and chopped the log to pieces, but didn't find the snake (Baughman and Morgan 1977).

1959—
A father and son were picking berries near the Duvall Cemetery, when they heard rocks rolling down a strip mine pile. Investigating, they found a huge snake they estimated as 35-40 feet long. The man and his wife also claimed to have seen it the next summer in the same area (Baughman 1987; Clearfield, Pennsylvania, *Progress*, May 21, 1975).

1971—
A man cutting paperwood near Paradise Furnace claimed to see a large snake (Baughman 1987).

1979—
A family driving near the Weavers Falls boat ramp ran over what looked like a log in the road. They stopped, then realized it was a huge snake that crawled rapidly away.

1980s—
When I first began investigating reports of this snake, Jon Baughman kindly printed a research request in the *Bulletin*. This elicited a very intriguing response from a woman who had seen a very large snake in the area, but didn't realize there was a history of sightings. Being familiar with snakes (she has kept a pet snake), she wasn't frightened and was able to give the following details of her encounter:

"As best I can remember I saw the snake between 1985 and 1989. My mother is from Broad Top Township where her family has lived for many years. I was driving to Everett on a road that would take me past Lake Groundhog, but still on the other side (south) of the mountain from Hopewell. As I was driving I saw a huge snake coming down off the hill to my left. I couldn't believe what I was seeing, so I slowed down watching it out of my side mirror. When I came to a stop (then & now there's not enough traffic on that road to worry about) the snake stopped dead behind me. I figured the snake was playing dead and got out of my car and stood beside it. The snake's head was in the brush on my side of the road and its tail clear across the road to the other side. I stood there for several minutes just looking. It didn't move an inch. I got so worried that it would still be playing dead for ages after I left and get run over so I stomped my foot beside its tail and the snake took off in the direction he'd been going.

"As to what the snake looked like, its head was hidden in the brush. I never saw it. As best I can describe the color, the snake was a very pale tan, maybe the color of pinewood. I don't remember any pattern, nor did I see his belly. I've seen some impressive black snakes in the area and it was nothing like them. It was thick around, like a boa or python."

2000—

In early August, three men were driving to work in the very early morning, planning to fill up their water cooler at a natural spring, when they saw what they initially thought was a log laying across S. R. 1025 (or Enid Mountain Road). They quickly realized it was a huge snake. They saw at least 17-foot of it covering the road (later established with a tape measure), but weren't sure how much more of it was already in the tall grass; they never saw the head. They said it was slate gray with white markings. (Saxton, Pennsylvania, *Broad Top Bulletin*, August 8, 2000)

Quebec

1. Saint-Jean-Baptiste

Canadian Serpent.—A late Montreal paper gives an account of a snake, first seen by a woman and some children, and afterwards by several men, which, when twisted round a tree 30 feet in height, reached from the bottom to the top of it, and *appeared* (for they did not measure him) to be as big round as a common water-pail. A horse, sex or seven neat cattle, and some sheep, are missing from the neighborhood, (a place called *Jean Baptiste*,) and of course he is charged with having carried them all off for his own eating. The Montreal editor does not seem to doubt the truth of the story as his information is derived from the owner of the ground in which his snakeship was seen, corroborated by half a dozen men, who had looked on while he glided through grass and over plowed land, to hide himself in a wood.—[We should be very loth to have such gentry lurking about our premises.]

Portsmouth, New Hampshire, *Journal*
July 29, 1826

Large Snake.—To the utter loss and mortification of that scientific paper, the Montreal *Herald*, and all honest cultivators of onions, we have to announce the non-existence of that astonishing reptile which occasioned so much uneasiness in the country, and so much ingenious speculation in town. The truth is that the whole matter was a hoax, and that all ideas with respect to a large serpent originated in the sagacious mind of an honest cultivator of L'Assomption, who had accidentally discovered an extensive plat of strawberries on his land; and, as the only means of avoiding all unnecessary visits from his neighbours, and of securing the entire booty to himself, fell upon the capital plan of conjuring up the large snake of which we had such an elegant description in the Morning *Herald*, followed up by one of the most luminous scientific speculations that any age or country has hitherto been favored with.—Mont. *Gaz.*

Portsmouth, New Hampshire, *Journal*
August 19, 1826

South Carolina

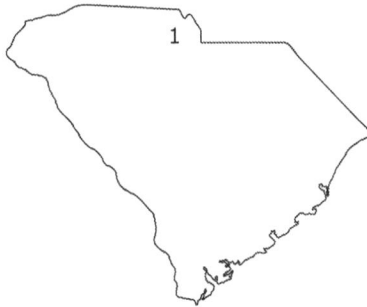

1. York, York County

From York and Yorkville:
"There is a 50-year-old myth of a tremendous snake that
makes its home on the lot on East Madison street next to the
municipal power house. Recently it was reported that a prominent citizen, while driving a car in broad daylight encountered
the snake crossing the road in the direction of the oil mill to which
it was going in search of a dinner of rats. The snake was alleged
to have been long enough to stretch entirely along the road with
its body six inches through, and the prominent citizen, afraid to
attempt to run over it stopped until it had passed by. Investigation, however, failed to develop the slightest corroboration from
the prominent citizen in question. It was the first time he had
ever heard of the big snake. Although the fiction has persisted

at least 50 years during all that time there has been found no reputable citizen who claims to have seen the reptile. Considering the age of the snake it ought to be a large one, and considering the imaginary status of it, there is no reasonable limit as to its size."

Gastonia, North Carolina, *Daily Gazette*
May 6, 1920

Tennessee

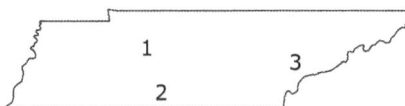

1. Triune, Williamson County
2. Pulaski, Giles County
3. Loudon, Loudon County

Serpent Thirty-Five or Forty-Five Feet in Length and Eight or Ten Inches Thick.
The Terror of Triune for Twenty-Five Years—
He is followed to his Den and will Probably be Captured.

(From the Republican *Banner*.)
Triune, Tenn., July 5, 1868.
Editor Banner: A most surprising circumstance occurred in this neighborhood last Friday, which has created intense excitement. Believing it may interest your readers, I have taken the trouble to collect the facts and visit the locality, and in order to put the truth of the narrative beyond cavil, you will find appended thereto the affidavits of the gentlemen before Esquire Page, of this place. I will premise the occurrence with a short history of what is known in this community as the "Big Snake."

About twenty-five years ago, as Mr. Vernon was returning from his workshop near Lindall's Mill, on the Franklin and Murfreesboro road, at the foot of a large hill, it being one of the range known as the Burke Hills, he saw the track of some immense serpent, leading toward a large thicket on the hill-side. It measured about nineteen inches across, and seemed to press into the dust, as if a great weight had pressed over it. Mr. Vernon collected the people of the neighborhood the next morning, and well-armed, they made a close search for the monster, but without success. About ten years afterwards his track was again seen by several men about a mile from the former place. Some five years later, as Mrs. Barnes was in her garden gathering vegetables, she saw the monster lying along a cabbage row, but was so badly frightened that she was unable to form even a conjecture as to his size. Her husband quickly ran to the garden when the alarm was given; but the snake had disappeared, leaving, however, his immense trail.

In the fall of the same year, as Mr. Allen was cutting a pole in a thicket of Mr. Boyd Barnes, with whom he was living, he heard in the thicket what he supposed to be a cow or horse, and the thicket being in his cornfield, he started in to run the creature out. What was his horror, on taking a few steps, to see the head of the veritable big snake, reared about eight feet high, and gazing upon him intently. Of course he fled at the top of his speed, and did not turn to see what course he would take. A few men on horseback went to the place, but could only find by the broken bushes the route he took.

These circumstances following close upon each other, greatly excited the neighborhood, and mothers became fearful of leaving their little ones alone, lest they should be swallowed by the terrible animal. Mr. Allen described him to be as thick as a large man's thigh, and said his head was about the size of a frying pan, and nearly flat.

Nothing more was seen or heard of the snake before or during the war. People hoped he had died in his den from old age. But it remained for Messrs. W. S. Robinson and J. W. Lisle, of this place, to catch the next glimpse of him, and it occurred in this wise: Last Friday Messrs. Robinson and Lisle were riding toward Nashville on the pike, and as they neared the gap of the Burke Knobs, they saw what they supposed to be a large pole stretching across the pike, not thinking of the improbability of a pole being in such a locality, as the pike has a large field on either side. To their surprise they saw the supposed pole, when they got within sixty or seventy-five paces of it, suddenly raise its head above the top of the fence on the west side of the pike, and commence moving slowly over it. Although the two gentlemen were badly frightened, they determined to see more of his snakeship, and as soon as he had got about one hundred yards into the field, they passed beyond him on the pike and quickly got to the top of the hill, some two hundred yards above him. The serpent did not seem to care for their proximity, but quietly pursued his way, which lay about half way down the hill-side through the field, until he came to the bushes in the woods. Here he stopped awhile, and lay very still and quiet until the gentlemen began to think he had put up for the day. But they waited patiently above, ready for a quick retreat should he start toward them. He remained there nearly two hours, until mid-day, when he raised his head full fifteen feet, looked around slowly, and resumed the same course through the woods, with head erect. This position he did not change for nearly half a mile, when they saw him dart suddenly forward with great velocity and strike toward the ground. When he again raised his head he had a hare in his mouth. He halted her a while, and lowering his head, they supposed he swallowed it, for after remaining down for fifteen or twenty minutes he raised his head and moved slowly forward.

The woods here becoming open and free from bushes, they had a fine view of his size and length, and becoming cooler and more possessed, they were able to come to some conclusions as to their course. Being both of venturous spirit, they determined to watch him to his den, and, if possible, effect his capture. Their determination being made, they, without a great deal of difficulty, kept in sight at a safe distance, until they passed what is known as the Morton Graveyard Hill, when the serpent passed through the gap beyond that, and directed his course in a westerly direction, along the North face of the hills. He was often hidden from their view by intervening thickets, but knowing of his previous appearance, they supposed he was going to his den in that direction, and thus, by getting a clear view beyond the place where he was hidden, they soon would be able to see him emerge.

When the ground was clear of obstructions, he traveled with his head near the earth, raising it as he approached a fence or bushes. Thus they followed him up near two and one-half miles, until he came to a large hill that overhangs the residence of Mr. Burke. This hill is very rocky and precipitous, and has numerous fox holes in its side. When about half way up this hill, the snake stopped in a perfectly open place, and after looking all round he gave forth a low, bellowing sound like the suppressed lowing of a cow, and then thrusting his head into a hole that was invisible to them, he slowly drew himself into it.

Mr. Robinson now remained on guard while Mr. Lisle, with the assistance of Mr. Thomas Burke, quickly aroused the neighborhood, and coming to the place, each with a large rock in his arms, they quickly stopped up the hole. It was about two feet in diameter, and seemed, from the place being worn smooth around, to have been occupied for some time. Mr. Burke had noticed the hole frequently, and supposed it to have been occupied by groundhogs.

All yesterday was occupied by the neighbors in constructing a long box, of heavy two-inch planks, one end of which was thrust into the hole, and rocks piled around, the other end crossed with iron bars, so as not to obstruct the light. A sliding door next to the hole, and a piece of mutton at the far end, attached to the door by triggers, gives them hope the monster may be caged. The box is fifty feet long and eighteen inches square, heavily bolted together. The gentlemen think the serpent is fully thirty-five or forty feet long, and about eight or ten inches in diameter.

Where he came from, or to what species he belongs, can only be left to conjecture. He was of a uniform dirty black color. A close watch will be kept on the box, and if their expectations are realized, you yourself may yet have a glimpse of this extraordinary creature. I will report what success they may meet with, and only hope it may be captured and carried to your city for the inspection of your savans.

Respectfully, J. L. Scales

State of Tennessee, Williamson Co., July 4, 1868.

Personally appearing before me, an acting Justice of the Peace for said county, Messrs. W. S. Robinson and J. W. Lisle, who made oath that the circumstances related in the communication of Mr. J. L. Scales to the Republican *Banner* are substantially true, as far as it relates to them, and I certify that I am personally acquainted with the above named gentlemen and know them to be gentlemen of veracity.

S. H. Page, J. P.

Janesville, Wisconsin, *Gazette*
July 17, 1868

The Terror of Triune.
More from the Monster of the Burke Knobs—He Smashes the Big Iron-barred Trap and Escapes—A Spectator Frightened into Convulsions and Driven Mad—His Enraged Snakeship Clears for Tall Timber.

Yesterday we published a long account from the Nashville *Banner*, of a big snake seen in the neighborhood of Triune, Tennessee, and the particulars of his being traced to his den, and the preparations made for his capture by the construction of a cage over his hole. From a gentleman who conversed with a number of people in Nashville yesterday, we learn that this story is perfectly true, and that many people living in the vicinity of his snakeship's quarters are terribly alarmed, and one man was driven mad. The following is a second communication to the *Banner* on the subject:

Triune, July 8, 1868.

I hasten, agreeably to my promise, to inform you of the success we met with in our attempts to capture the big snake. The box was watched at various times through the day, but nothing was seen or heard of him until late Monday evening, when a Mr. Palmore, who lives near, visited it, and familiarity with danger lessens fear—having been there so often and seeing nothing—he carelessly threw himself before the open end, placed his face to it, and looked in, but only looked, for there, within three feet of his face, was a pair of blazing eyes, like coals of fire, glaring at him. He didn't stop to see how large the head was, but threw himself down the hill with such head-long speed as to catch a series of falls, in one of which he very seriously injured his head. Had it not been for the accidental visit of Mr. Palmore he would very quickly have taken the bait.

But it seems he got a scent of it, for, as the sequel shows, he soon came back.

This morning Mr. Thomas Burke got up when it was fairly light, and while putting on his clothes he heard the most singular noise—a low deep bellowing quickly given and then a crashing sound. He ran to the door and looked out to see the cause. It was quickly revealed to him, for, after a tremendous crash as of a breaking tree, in the direction of the trap, he saw the awful monster coming down the hill with the most frightful leaps and bounds, writhing and twisting his huge body as if in great pain. He passed across the road about two hundred yards north of Mr. Burkes', sweeping down the fences and other obstacles in his pathway. But this Mr. Burke did not see, for, with the first glance at the serpent, consciousness left him, and he remained for a time in convulsions from the fright.

It was late in the day before any one would venture near the trap, but at last some went up, and they found it lying with the end still in the hole, but the other riven as if struck by lightning, some of the bars of iron being found full fifty steps off. Thus the capture is a failure, and the serpent is again at large, and the country excited to the last degree. Many speak of leaving the neighborhood of the hills, and serious fears are entertained of the effect on the mind of Mr. Burke. No one supposed, for a moment, he would be able to break the immense cage prepared for him—it would have held a lion. But we miscalculated his power, and he is free.—*Cin. Enquirer.*

Fort Wayne, IN, Daily Gazette
July 16, 1868

The Great Tennessee Snake
The Monster Shot Dead by His Pursuers—His Length, Thickness and General Appearances—The Skin to be Exhibited Throughout the Country.

Triune, Tenn., July 14, 1868.
To the Editor of the Nashville *Banner.*
My letters to you on the subject of the "big snake" seem to have excited more attention than I dreamed of, and I am sorry

that I wrote at all, especially as so many express doubts as to the correctness of my statements. Even you, yourself, seem to think that I am only trying the gullibility of the public. I would not now come to you again, even with a sworn statement, did I not have proof that will convince the most skeptical, all burlesque and sensational telegrams to the contrary notwithstanding. "Truth is stranger than fiction," and if a man only believes what he sees, how little of history would be recorded.

But I am even able to offer, to a doubting public, that proof, namely: The monster himself *in propria persona*—for we have him, the fiery snake that has created such an uproar. However, I will proceed with my narrative, and give you the plain, unvarnished statement of facts:

I wrote you on the 8th instant, that the serpent had broken his box and cleared for the woods. This created a great deal of consternation in the community, and all lay low until towards evening, when the neighbors got together and devised some means to put an end to him. On a close examination of the box it was discovered they had put a defective plank in it, it being windshaken from one end to the other, and thus, when the captive began to surge, he soon started this plank and was then able to break out.

The hole in the ground was now securely stopped, and every one present determined to make it their business to hunt him up and destroy him. They felt it necessary for their safety. Strong bands well armed went together and scoured the woods, beat the thickets and searched closely everywhere until they began to despair of success. But, when they least looked for it, fortune smiled upon their efforts. Some mile and a half or two miles from the den lives a Mr. Isaac Neily and a Mr. Irvine, close neighbors. Yesterday morning Mr. Neily walked over to Mr. Irvine's carrying his gun, an Enfield rifle. Finding Mr. Irvine in his field, nearly at the top of the hill, at work, he called to him, and Mr. Irvine and he seated themselves on the fence, with their guns (for no man now went without one) near them. They had talked some time when they became aware of a peculiar musty odor pervading the atmosphere. They had felt it at first, but it did not excite their attention until it became sickening in its power. Mr. Neily looked around unconsciously to see what could produce it, and there, within sixty yards of them and just above, lay the serpent, the object of their dread.—Their first impulse was to run, but after they had involuntary [sic] jumped from the fence they thought better of it, and, like true men, as they are, they determined to fire on it. They laid their guns carefully on the fence, the snake never moving, and taking deliberate aim, fired at his head. At the cracking of the guns an awful slashing and crashing was heard, and believing the creature was upon them, they threw away their guns and fled for dear life, nor did they stop until they reached the house.

The country was quickly alarmed, and after a good deal of careful reconnoitering, the spot was approached, to find the serpent gone, but leaving behind him such a great trail of blood they knew he was badly wounded. Without difficulty they followed his trail, and did not proceed more than two hundred yards down beside the fence before they espied him, perfectly limp and dead. But it took several shots from a distance to convince them of this before they dared approach.

The struggle when the monster was shot must have been terrific, for large bushes were bent to the earth, and the ground was torn up, where he had lashed it with his tail. Great credit is due to the intrepidity of the gentlemen who shot him, for few would have dared it.

Mr. Irvine quietly brought a pair of mules to drag him off, but when the mules caught sight of the huge animal they became so frightened nothing could hold them, and they broke ad tore away like mad. By this time a great many people had collected, and they tied a chain around his neck with a running noose, and dragged him through the field to the big road. They now found a couple of strong horses gentle enough, and they with these determined to carry him to Nolensville, it being the nearest point on the pike. Hearing of these circumstances, I, in company with Mr. Ben Johnson, of this place, started for the scene of action. When near Mr. Bittock's, we met with Esquire Greene going to the same point. We had not gone far before we hove in sight, and here we met with a very serious accident, for no sooner had the horses caught sight of the huge body of the serpent than they began to plunge at such a fearful rate as to throw all three of us, Mr. Johnson and myself sustaining no injury, but Esquire Greene received a severe dislocation of the ankle, that will probably lame him for life.

However, we soon reached Nolensville, and the question arose, what will we do with it? We were anxious to take it to Nashville, just as it lay, but the carcass would, we knew, become very offensive from the great heat of the weather. So at last it was agreed to skin him, and after stuffing and drying take him down to the city for inspection. I took a good survey of him and found his length was not so great as we had imagined, for he was just twenty-nine feet eight inches from tip to tip. His greatest circumference was thirty and a half inches, and he was just seventeen inches around the neck, or rather just behind the head. About ten feet of his tail had a spinous ridge like that of an alligator. His head was the most formidable part about him, and was truly terrible to behold, even when dead. It was fully twenty inches across, and was armed with the most formidable set of sharp small teeth, with four huge tusks. It had a hole in it just under the right eye, from the minnie ball, and a dent in the neck from the glancing ball of Mr. Irvine. Although he seemed, at a short distance, to be of a uniform dark color, on close inspection, I discovered he was spotted like a rattlesnake, in diamond spots, and his belly was a bright bluish-black color. We cut him across in the middle, and skinned him both ways, thus preserving the head whole. It was then well salted and lined, and the ends were well sewed together; and then, by inserting Mr. Simpson's bellows, we inflated the monster to its full size.

We will let it dry in this fix, and then stuff it with sawdust. And now comes the best of all.

There are so many shares in it, its captors have agreed to make a present of it to Andrew Johnson. This determination was arrived at in the face of a very liberal offer sent for it by Messrs. Wall & Handy, of Franklin. Now, in order to pay expenses there, as soon as it is in traveling condition, it will be exhibited in all the large cities between here and Washington, commencing in Nashville the first of next week.

J. L. Scales

Janesville, WI, *Gazette*
July 22, 1868

The Pulaski (Tenn.) *Citizen,* comes to the front with the following snake tale:

In the Sixth District, a few days since, a well-known farmer, while walking through his cornfield, was attacked by a monster snake. The monster made his approach with his head three feet above ground. The farmer being unprepared for the attack, naturally gave back, but the snake pressed upon him so close that he frantically struck at him with his fist, having nothing else to fight with. He finally succeeded, however, in procuring a stone and killing the monster, which measured eleven feet long and fourteen inches around the largest part of the body. Its eyes were very large and white. About a foot of its tale was plaited like a whip, and it was recognized by old settlers as a "horse-whip snake."

Galveston, Texas, *Daily News*

July 28, 1875

Giant Snake at Large in Country
Tennesseans offer $100 reward for Reptile and Loudon Brames Arm Is Carrying Off Animals
Shot by Farmer His Snakeship is Declared to Have Torn Up Sapling in Getting Away from the Scene

Knoxville, Tenn—A giant snake said to measure at least 25 feet in length, 8 inches in diameter, and 24 inches in circumference is terrorizing the people of Loudon and that section and a reward of $100 has been offered for its capture dead or alive.

Several posses of young Loudon braves have formed for the purpose of capturing the reptile, but as yet without result. It has been seen off and on for the past 28 years, but not until this year has it caused any serious alarm. Recently it left its track in the field of W. T. Huff, resembling in size that made by a log dragged by lumber wheels. A woman saw the snake which actually ran over a fence, tearing it down, being unable to get through the cracks. She thought it was the devil but for the fact that it ran from her instead of toward her. A few days later, a lamb belonging to Eff Huff was caught and killed by the reptile.

Ed McQueen, a prominent farmer, has stopped his children from going after the cows in the section where the snake is said to reside. John McDonnell, a farmer, is the only man who has got a shot at the snake. He saw it going toward the river and then drinking at the edge of the water, and from his boat he fired a charge of small shot into the snake from his Winchester.

The reptile, in a frenzy from the pain, tore up saplings in getting away. No one will now make the trip in Loudon over the road leading to Pond Creek, Sweetwater, and Kingston.

Men from Loudon on horseback and with their guns are doing picket duty. Arthur Rodgers and Ned Cassady are armed with guns which shoot steel bullets. Women and children have barricaded themselves in their homes and the least noise causes them to fear and they cry, "The snake!"

Dick Ferguson is now organizing a posse to go after the monster, which is believed to reside in a cave on the bluff overlooking the river on the McQueen farm. There is no known species of snake growing to such size in this section.

Eureka, Utah, *Reporter*
August 28, 1908

Texas

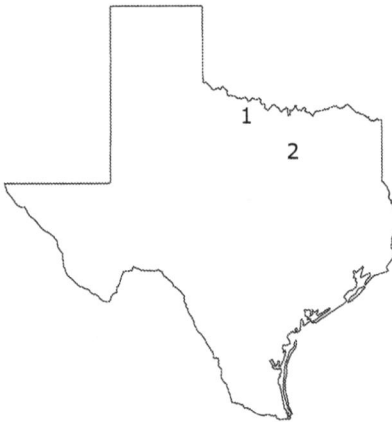

1. Wichita Falls, Wichita County
2. Fort Worth, Tarrant-Denton Counties

Trucker Reports Seeing Huge Snake

Wichita Falls, June 22. (AP)—Amateur snake experts were engaged in speculation here today after James H. Hankins, driver for a lumber company, reported he saw a snake, about 20 feet long and 10 inches in diameter on the Wichita Falls-Seymour highway.

Rankins said when he attempted to run over the snake it coiled and reared as high as the cab window of his truck. He said it definitely was not [a] rattlesnake.

> Big Spring, Texas, *Daily Herald*
> June 23, 1946

The Great Snake Story

The Wichita Falls *Record-News* reports that the gigantic reptile which kept the people of that community on tenterhooks for weeks has finally been exposed as the hoax of a group of teen-age youths in search of a little good clean fun.

It was only a dummy, made of tow-sacks sewed together and filled with cottonseed, the whole painted black to look business-like. It made numerous appearances along the highways and byways of the vicinity, and scared a good many people out of their wits.

Posses were organized to hunt down the monster. One man is reported to have offered $1,000 for the snake alive. Residents were made jittery by the wild rumors floating about, and the whole countryside was disturbed.

The Wichita paper says the boys had no intention of creating such excitement. They merely thought it would be a good joke. Their identity, the paper says, may never be known. The dummy has been destroyed. The show is over.

Hoaxes of this nature, it seems to us, are rarer nowadays than in the good old days. This particular hoax was different from most, in that there actually was a dummy snake, whereas most hoaxes are born and thrive on nothing more substantial than idle rumor.

The snake hoax was funny to its perpetrators, but hardly a laughing matter to those who were victimized by it. Fortunately none of the victims died of heart failure, as they might well have.

Those who were actually exposed to a sight of the snake probably are no more resentful than those who were duped into offering rewards for its capture, said to have reached a total of $5,000, or to the owners of "snake dogs" which were offered as trackers-down of the putative python.

At any rate, the big snake scare around Wichita Falls is over—unless there are some who persist in believing that the story of the hoax is itself a hoax. Error sometimes persists in the face of facts.

Abilene, Texas, *Reporter-News*
July 17, 1946

Tip Touches Off New Hunt For Python Pete

Fort Worth, Tex., Oct. 1—(AP) A Dallas man who said he saw a big snake as big as a stovepipe swimming in the Trinity river near Fort Worth today touched off another search for Python Peter.

Zoo officials and officers hurried to the scene in the hope of finding the 18-foot python that disappeared from the zoo two weeks ago.

David W. Smith, 59, said he saw the snake from a highway bridge.

"I was standing on the bridge when I spotted this big brown snake 15 to 20 feet long and as big as a stove pipe swimming toward a bend in the river," said Smith.

"I watched him until he rounded the bend which was about 300 feet away."

Smith said he had not read of the python's escape but had heard something about a snake being loose.

Bridgeport, Connecticut, *Telegram*
October 2, 1954

Regarding this last story, Python Pete was finally discovered 125 feet from his cage by October 4, 1954. His keeper stated that the snake was probably there all the time, so sightings of the snake elsewhere were likely mistaken.

Utah

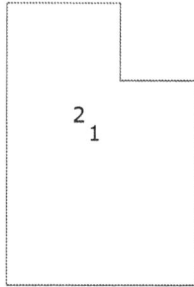

1. Utah County
2. Salt Lake County

The Champion Snake Story.—Some of our readers may remember hearing a rumor, afloat in this region some years since, about a huge serpent that had been seen by an Indian somewhere in the mountains in this or Utah county. Such a story has been in circulation once or twice within the last ten or twelve years. A letter, written by a gentleman at Spring Lake Villa, Utah county, printed in the *Deseret Weekly News*, of September 3rd, 1862, contains the following:—

"Last week an Indian brave and squaws went over the first mountain east of here to pick berries, and next day returned very much frightened, and for some time refused to tell what his trouble was; but finally says he went over on the other side of the mountain, and then, after walking about a while came upon the trail of something appearing as though timber had been dragged along. This he crossed, and upon going a little higher, saw the head of some living thing peeping over a rock at him, which much frightened him, as he had never seen anything like it; but before he could run, a huge snake, as thick through as a man's body, and from ten to twenty feet long, sprang towards him. He dodged, and the huge reptile went clear over his head, but gathered himself up again, as the Indian started the other way, and again jumped. Another dodge saved him again, when he scrambled over some rocks and up the mountain, getting not only clear of his snakeship, but nearly frightened out of his senses. He says the snake had horns that curved back over his head, and that he would raise his head to the height of a man."

The writer of the above letter says the brave's story was generally believed by those who were acquainted with him, but like the stories told of the Bear Lake monster, and also of the Utah Lake serpents, it has been regarded as a myth by the general public; but if a narration made to us a few hours ago, is to be relied on, the general incredulity in regard to these Utah monsters may not be so strong hereafter as it has been.

This morning a well known resident of this city and Territory, named Edward R. Walker, now of the 16th Ward of this city, a stout, strong man, about thirty-four years of age, who has been used to mountain life, and apparently as sound in mind as strong in body, called upon us and made the following extraordinary statement, which we have no doubt our brethren of the quill will stamp as the champion snake story of the entire country for this season. On the 16th instant Mr. Walker, his brother Sylvester, and their cousin John Coon, were felling timber for Mr. Standish's mill, in the right hand fork of Coon's Canyon, about three quarters of a mile from the Point of the mountain west from this city, on the high peak, south of Black Rock. Between ten and eleven in the morning a deer ran by where they were working, and our informant snatched up a Sharp's rifle, and started in pursuit. When he had continued the chase for about a mile, due north, he was startled by a loud, shrill whistle and hiss, which he at first thought might be a signal from an Indian. He came to a halt and looked about him, and heard the noise of rocks rattling south-east from where he stood. He turned, when to his horror he saw approaching him, at a very rapid rate, a serpent, which he judged was between thirty and forty feet long, and about ten inches through the body. The reptile's head was raised fully six feet from the ground, and his jaws were open fifteen or eighteen inches wide with large fangs growing from both upper and lower jaw. Walker was almost petrified with fear, but the hope of saving his own life made him start to run. The serpent, however, was too quick for him, and jumped at and knocked him down, striking him on the left shoulder just below the shoulder blade, going over him and down the mountain to the south-west for a short distance, when he turned and pursued Walker, who had risen and with a speed inspired by the deadliest fear was making his way to the top of the ridge. Unfortunately for our informant he stumbled, and immediately he felt the weight of his monstrous pursuer gliding over his body. He gave himself up for lost then, but it seems hard to tell which was most frightened, the man or the snake, for the latter did not seem disposed to run the risk of a contest, but after gliding across the body of the prostrate man, he slid off at a tremendous rate towards the ridge of the mountain and across it to the east side. Walker rose and watched his movements, and says that after crossing tot he east side, the snake turned and recrossed to the west side and went down the mountain a few yards, and then twined himself around a large mahogany tree, where he remained waving his head to and fro, flapping his tail on the rocks, and whistling and hissing defiance. That was the last he saw of him, for he made his way back to his companions as quickly as he could. They wanted to return and hunt for the serpent, but the hero of the adventure was too weak from fright and excitement to do so, but says they intend to go on an expedition to hunt that snake in a very short time.

The color of the reptile was yellow, with a black mark like a half moon on each side of his eyes; he had a beard or fuzz round his mouth, and what appeared to our informant to be a crown shaped mass on the top of his head. The latter was about six inches high, and varied in color, being green, blue, white, yellow, and red. The head of the creature was about as large as that of a full grown bull dog, and in shape between that of a bull dog and monkey. His body was covered with scales, six or eight inches long.

Mr. Walker says he has been used to mountain life for years, and never was afraid of anything; but nothing could persuade him to go alone again, into the right hand fork of Coon's Kanyon.

Such is the story of Mr. Walker. While he was telling it we asked him to allow us to feel his pulse, but it was perfectly natural, neither it nor any glare in the eye indicating the least degree of mental aberration. At the risk of offending him, his statement being a pretty strong one, we asked him if he had been drinking and had got "snakes in his boots." at the time of the "adventure." He assured us he was perfectly sober, that every word he told was true, and in proof showed the bruise on his shoulder caused by the blow he received from the serpent.

Deseret, Utah, *News*
July 30, 1873

Vermont

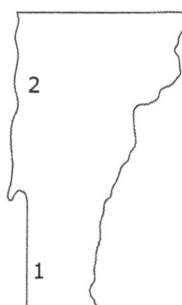

1. West Mountains, probably refers to the Taconic Mountain range that extends into the southwest corner of the state.
2. Richmond (Chittenden County)

"Sea-Serpent" in Vermont

The largest snake ever heard of in this part of the world has been seen for some months about a hedge on the east side of our West Mountains; and last Sunday Mr. T. Owsley saw him and describes him as follows:—He is as large as a common stove-pipe and about 12 feet long, as near as he could judge, but he dare not attack him. He had rather a venomous look. His color is a dark brown. The Green Mountaineers are intending to make a grand *sortie* and capture his snakeship.

Mr. Wood, Telegraph operator of the Vermont and Canada line here, says the above "snake story" is well authenticated, and pledges the veracity of the telegraph on the existence of the identical monster.— Troy *Post.*

Prairie Du Chien, Wisconsin, *Patriot*
August 30, 1848

Big Snakes in Vermont

The St. Albans (Vt.) *Messenger* of Saturday says: "Some six years ago a menagerie of wild animals came through Richmond, and broke down a bridge about a mile east of the village. At the time numerous reports were afloat that they had killed some of their large snakes and some had got away. Within a year or two the boys have brought reports of seeing very large snakes in the vicinity of the accident. A few days ago William Fields, a truthful, sober man, was on the farm of Jonathan Fay close by the railroad, and was startled by seeing two large snakes of the color of boa-constrictors. He says they were full six feet long, and when they saw him they raised their heads and opened their mouths and darted out their tongues. He says they were as large as his arm. Mr. Fields did not stop to interview them. Dr. Bromley, a few days ago, when driving through the Jonathan Fay farm, thought he saw a rail across the road. This proved to be a huge snake which moved out of the road. The Doctor thinks he was 10 feet long. Mr. Fay's men have found a snake-skin some 10 or 12 feet long, supposed to be shed by a monster snake. Making all due allowance necessary for exaggeration, it is quite certain that there are two monster reptiles, and probably three—as the skin found and the one seen by the Doctor correspond."

New York, New York, *Times*
July 22, 1878

Virginia

1. Lorraine, Henrico County

His Overalls Nipped by an Anaconda

Dr. Wendlinger came in town from up the Richmond and Alleghany railroad with a dead snake that was a regular monster. It was fifteen feet long and as large in proportion, and was evidently an anaconda which had probably escaped from some of the many traveling ten cent shows which have lately frequented this part of the world. His snakeship was left at the cafe of Mr. Gus E. Delaware, where a number of people gazed in silent awe upon the deceased gigantic reptile.

It was killed at Lorraine, on the Richmond and Alleghany railroad, about six miles from Richmond, by Mr. Lane, section foreman on the road. It is stated that it crawled out of the woods and silently approached from behind a negro boy, who was working near the track. It seized the overalls which the negro had on in its mouth and began chewing on them. Some one called to the negro to look behind him, and as he did so he saw the snake and fell over paralyzed with fear. Mr. Lane then seized an ax and killed the creature.

Whether the reptile intended to get a firmer hold on the negro's clothing and then, throwing him, and after crushing him to jelly eating him, or whether it was a tame snake that approached the boy only with friendly intent will never be known. An engineer of the road says that he saw the monster about two years ago and reported the fact.—Richmond *Times*.

Reno, Nevada, *Daily Nevada State Journal*
December 30, 1891

West Virginia

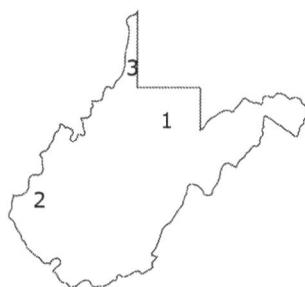

1. Grafton, Taylor County
2. Milton, Cabell County
3. Wheeling, Ohio County

A Fourteen Foot Snake.
He Makes His Appearance in West Virginia—
First Seen Thirty Years Ago.

A monster snake was discovered at Knottville, two miles from Grafton, W. Va. David Baker, who saw it, says that he, with others, encountered the snake in the woods. He described it as about fourteen feet long and of a dirty clay color. It was as thick as a man's waist and because of its size its head was raised about two feet above the ground. When it was first seen one of the men went to a house nearby and secured a gun while the others watched, but on his return the snake had disappeared. The monster has been seen before in the same neighborhood. Ten years ago he was seen by three children, who state that as he crawled through the long grass of the meadow he left a broad track behind him. He made his appearance again two years ago.

All of these different tales agree as to the appearance of the monster except that when he was first seen, thirty years ago, he was but ten feet long. The inhabitants along the swamp express their fixed determination to capture the beast dead or alive. They have found his den and will watch until the snake makes his appearance.

Marchfield, Wisconsin, *Times and Gazette*
April 26, 1884

Another "Big Snake" Story.

Huntington, W. Va., Oct. 4.—A letter just received by the *Republican* from Milton, says a monster serpent has been discovered in a cave on Big Two Mile Creek. It is described as being forty feet long, with black and yellow stripes running lengthwise of its body. The head and neck are black, and the tail is of a dull brown color. This monster was discovered by Melzer Braley, about twenty-three years of age, while passing "The Buzzard Den." His attention was first attracted by the bleating of a sheep. Going close to the mouth of the cave he found that the sound issued from it, and turning the corner of a large boulder he saw the head and a portion of the body of a reptile as it was in the act of swallowing the sheep, the hindquarters of the animal having already disappeared down the snake's throat. Running up the hill, Braley rolled a large stone upon the reptile, causing it to disgorge the partially swallowed sheep, and then with loud hissing the snake glided from the cave. Braley fled in terror to the home of his brother, two miles away. Several men soon gathered near the cave, but the snake was not seen again, although its hissing could be heard coming from within the cave.

Decatur, Illinois, *Saturday Herald*
October 9, 1886

There is a story going the rounds about a monster snake which is supposed to have been seen near Mount de Chantal Academy, near Wheeling, at various times recently. It is described as being about twenty feet long and as thick as a man's body.

San Antonio, Texas, *Daily Light*
September 7, 1889

Wisconsin

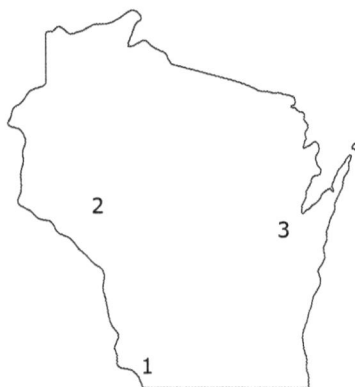

1. Specht's Ferry, across river from Potosi, Grant County
2. Eau Claire County
3. Oneida, Outagamie County

Beats the Record
Monster Snake Said to Have Killed Cows in Wisconsin.

Dubuque, Ia., Aug. 3.—Nicholas Premier, a farmer living near Specht's Ferry, in Wisconsin, some miles above here, was in the city yesterday and tells the most wonderful snake story ever heard in this region.

Premier says a monster serpent appeared in his pasture and killed two of his cows by winding itself around them. He saw one of his cows in the folds of the snake, but was afraid to interfere and ran away. Returning later he found the cow dead and a large hole in her flank.

The same thing was repeated next day, and a party was organized to hunt the monster, but when they came upon it they were also frightened and allowed it to escape. One man insists that he saw it swallow a large calf.

Premier describes the snake as about twenty-five feet long and three feet in circumference in the middle. The folds of the reptile left deep creases in the animals killed. It is supposed the snake has escaped from some menagerie.

Premier is well known here and his story is believed, incredible as it appears.

> Centralia, Wisconsin, *Enterprise and Tribune*
> August 8, 1896

Huge Snake is Reported
Resident of R. F. D. No. 2 Tells of Mammoth Serpent Seen on His Farm.

The following is received under date of June 20:

"Last evening before the storm, on my premises, a large snake, in my woods, near a spring, nearly reached the entire distance between two trees—which measured afterwards were found to be some twenty feet apart.

"About two weeks ago I lost a spring pig, which was running in the pasture, and have lost three more since that time.

"Large storm came up and I was forced to leave and have found no trace of the snake thus far.

"Respectfully yours,

"H. M. Arnsdorf.

In the absence of the snake editor who became indisposed in attempting to handle this case, the political reporter, who considered himself as next in line for duty on account of his familiarity with octopuses, questioned Hon. Wm. Cernahan of the town of Union today in regard to the big snake above referred to

as having invaded the Eden of Eau Claire Co. contrary to the peace
and prosperity thereof.

Mr. Cernahan said that he had not seen the snake, but as it
was a great growing season in spite of the cool weather, and
everything in the town of Union was three times as large this season as in previous years, he was open to conviction in regard to
the story of the monstrous reptile aforesaid. He had just returned
from that part of the county and would have returned sooner had
he been informed of the presence of the snake.

> Eau Claire, Wisconsin, *Daily Telegram*
> July 1, 1904

In September of 1981, a Menomonie farmer found 8-inch wide "slithering marks" on his property, which he and sheriff's deputies decided
were the tracks of a 10-foot long snake, probably a python. No snake was
actually seen, but a State Trooper passed along the rumor that a "reptile
truck" was reported stalled along Interstate 94 some time before. (Elyria,
Ohio, *Chronicle Telegram*)

Giant Rattlesnakes
by State or Province

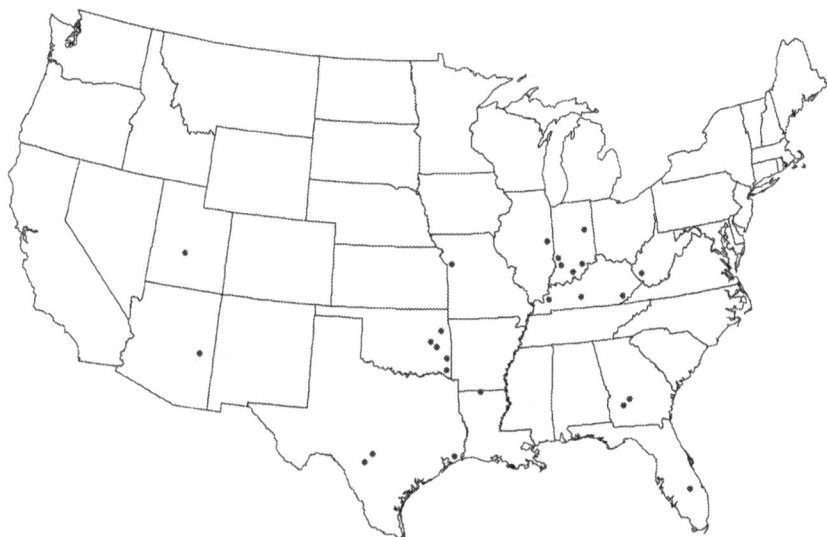

Reports of giant rattlesnakes go back at least to the 1700s. In a brief review, Klauber (1984) mentioned accounts of seventeen-foot long rattlesnakes from 1714, eighteen-footers from 1750, and a twenty-two-foot long rattler described in 1753. Unfortunately, he doesn't say where he found these accounts, so there is no way to determine credibility. Many colonial-period stories of oversized snakes and other strange beasts were told to distinguish and boost the North American fauna in comparison to their European counterparts.

As we noted earlier, there is some reason to suspect that *Crotalus adamanteus*, and possibly some of the other rattlesnakes, reached larger sizes in years past, but stories of 14-foot plus rattlesnakes just can't be taken at face value. Besides witness exaggeration, Klauber noted a trick where some claimants to substituted python or boa skins as "proof" of giant rattlesnakes. In those cases, of course, the head and rattles were always missing.

Another trick was to stretch a skin. Dobie (1982) didn't think a rattlesnake skin would stretch much more than an inch and a half per foot. I suspect that's true, as any further stretching should be very noticable, with the scales separating too far. But, I haven't seen this tested, so experimentation may prove otherwise. Different preservation techniques will affect how much a skin is able to stretch.

A bit outside our target area, but close enough to be of interest, comes a tale from Baja California, Mexico (Klauber 1984):

"For a number of years, we have been hearing stories about a very large rattlesnake said to live in some caves a few miles from our ranch. We have always listened to them with a smile and forgotten them. Last spring, a man who has worked for us a long time and whom we know to be truthful and not overexcitable, came and said he had seen the big rattlesnake. He said it was immense. He got terribly frightened and ran away. Now, three days ago, his son was gathering wild honey (the bees live in caves) when he saw the big snake. He had camped and was making a fire when he noticed his horse was excited. He looked and there, about twenty steps away, came the snake toward him. He ran to the horse and mounted and sat watching. The campfire was between him and the snake. The snake came on but when it neared the fire it raised its head four or five feet off the ground and hissed with such force it sounded like a bull. Then, it turned, went around the fire and down a draw, but kept its head up, looking back at him and hissing. The boy says it was no less than twenty-five feet long, and he thinks it might have been thirty feet. Its rattles were as wide as his three fingers and about a foot long. The boy was terribly scared, horrified, and ran home at once. Next morning he came here to tell us. We have always told him that, if they saw the big snake at any time, we wanted to see it. So my two sons went with him. They hunted all day but could not find any trace of the snake."

Arizona

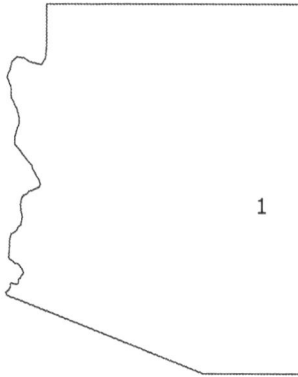

1. White Mountains

Rattlesnakes in Arizona (excerpt)

Tom Ewing of San Francisco, who erected several quartz mills in Arizona, was driving along one day when his progress was barred at an abrupt turn by a monster rattlesnake. His horse became panic-stricken, and as he was unarmed he was forced to turn around and seek assistance at the nearest station. Several men came out with shot guns, and after a fight, which came near proving fatal to one of them, the venomous reptile was killed. It stretched clear across the road, a distance of 14 feet 3 inches. Judging from the number of rattles, his age could not have been less than 42 years. I do not vouch for this story, but there are men in Arizona who claim to have seen the snake after he was killed. It is one of the traditions of the Territory.

Fitchburg, Massachusetts, *Sentinel*
Sept. 30, 1879

A Snake With 103 Rattles
Monster Reptile Killed in the White Mountains of Arizona Could Be Heard a Mile Away.

Tucson, Ariz.—The noisiest and largest rattlesnake ever killed in the United States was shot recently by Joseph Sponselor who lives at the sawmill on the White Mountain Indian reservation. He claims the rattler sported 103 rattles, was as big around as a man's leg above the knee, and the sound of its rattles could be heard a mile away. According to experts the rattler must have been about 300 years old.

One evening Sponselor heard a peculiar noise which he could not identify. Taking his shotgun he went in the direction from which the odd sound appeared to come to investigate.

He walked three-quarters of a mile before he came upon the snake. Without stopping to see why the snake was shaking its rattles so angrily he banged away and killed the monster.

The snake had a veritable cluster of rattles a foot long. The reptile was taken by Sponselor to the trading post where it was placed in alcohol and may now be seen. Now Sponselor is looking for the rattler's mate, for it is said they always travel in pairs.

Van Nuys, California, *News and Call*
February 9, 1912

Florida

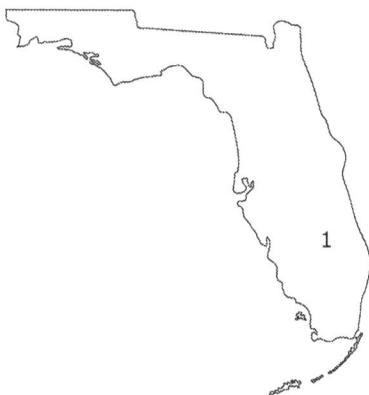

1. Fort Drum, Okeechobee County

We are informed by Mr. Long of Brevard County, Fla., that
while driving his ox-team near Fort Drum, in that county, his oxen
shied and ran out of the road. Seeing something raise its head and a
movement in the grass, Mr. Long, after stopping his team, went back
to see what it was. Upon approaching the object he heard a great
rustling and rattling, which convinced him that it was a rattlesnake,
but he could not see it, because of the palmetto and high grass, until
it threw itself into it coil and stood nearly as high as himself. He was
almost dumbfounded at seeing such a monster and hastily retreated,
but soon summoning up courage, he advanced near enough to be
within reach of the reptile with his long cow-whip, which he know

how to handle. With this weapon he opened the conflict, which lasted 15 minutes, Mr. Long keeping out of reach of the snake, but still near enough to strike it with his cow-whip, which was about 18 feet long. Finally Mr. Long began to feel sick and weak from the excitement, as well as from the musk emitted from the snake, and, putting in two or three rapid strokes with his whip, he retreated toward his cart, but fainted before he reached it. Upon coming to his senses again, he found that he had killed the snake. Mr. Long had no means of measuring its length but by his cow-whip, which was 18 feet long, and the snake lacked about two and a half feet of being as long as the whip. It had 38 rattles and a button. He says that it "was as large around as a big blue bucket." Mr. Long is one of the most reliable men of his section.—*Fort Read Crescent.*

Decatur, Illinois, *Daily Review*
November 3, 1878

Georgia

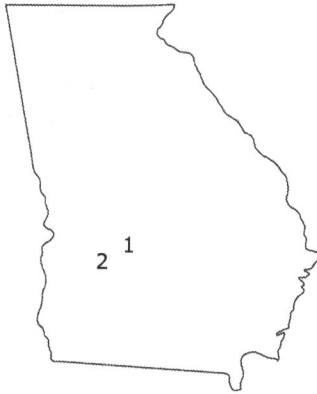

1. Dooly County
2. Lee County

The following we find in the Albany *Advertiser*: Mr. Dock Hall, "judge of Dooly county," says that he killed a rattlesnake last Friday evening near the bridge, that measured 14 feet 9 inches, had 43 rattles and tusks 1 1/2 inches long. He will have them at the Albany fair. He also has the hide, which holds 2 3/4 bushels meal bran, and has a chicken that weighs 48 pounds.

Atlanta, Georgia, *Daily Constitution*
May 9, 1878

The King of Rattlesnakes

The largest rattlesnake ever seen in Georgia was killed recently in Lee County upon the plantation of Secretary of State General Phil Cook.

The news comes, through Phil Cook, Jr., a son of the general. The snake has terrorized the neighborhood for years, and its death was the occasion of a jubilee celebration. The snake, by actual measurement, was a little over eleven feet long. It had nineteen rattles and a button.

The snake has been hunted for years, and traps innumerable have been devised for his capture. His den is in an impenetrable section of the Kinchafoonee swamp. Near this is a cypress pond, and between the swamps and the pond is the road. Hundreds of times his track has been seen across this road. People that had not seen it were loath to believe the stories told about it; but the truth finally became established and the Lee County rattlesnake became famous from the Atlantic to the Pacific. Every year the story is reprinted, with timely variations, to go the round of the press. The snake has swallowed young pigs, chickens, rabbits and other small animals, and was dreaded by the colored people like a ghost. It was difficult to induce them to travel the road between the pond and the swamp at night.

The snake was found across the path near the pond by Mr. Phil Cook. Without disturbing it, Mr. Cook went off for help, returning with three or four colored men armed with hoes and clubs.

Stealing up near the snake, the men fell upon it with the hoes and clubs and finally killed it. It was cut open, and in it was found a full grown buck rabbit. This probably accounts for the dormant and comparatively helpless condition of the snake and the ease with which he was dispatched.

The snake was then thrown across the shoulders of one of the colored men and carried to the house. Though he was a stalwart, muscular man, he staggered under the load. General Phil Cook says it was the largest rattlesnake he ever saw or heard of.

Al any rate, this is the end of the famous Lee County monster.— Atlanta *Constitution*.

Lafayette, Louisiana, *Advertiser*,
December 13, 1890.

Illinois

1

1. Ramsay, somewhere near Danville, Vermilion County

Snake Story—
Probably the Largest of its kind on Record.

The following snake story was told to us by a gentleman in whom we have every confidence, and we hope our readers will not reflect upon his veracity by doubting its accuracy, although we will not ourselves vouch for it further than as we said before, that our informant has too much respect for truth to "endanger himself in those sulphurous flames" into which Hamlet, and doubtless several others since, have been commited, by relating to us a miserable canard. He was called upon the latter part of last week to visit St. Louis, and on his return home stopped at a little station known as Ramsay, about twenty miles west of Terre

Haute, on the Terre Haute and Alton Railroad. The second day
after his arrival in the village, a man who had been hunting
brought in the news that he had discovered s huge snake, but on
discharging his gun at it his snakeship retreated to his abode,
within a cliff of rocks.

This statement confirmed what had previously been related
by an old settler in the village, which was, that while gunning
near this identical spot, thirty years ago, he also surprised an
enormously large snake, which at approach had sank away in the
same manner. The latter story had been repeated so often that
the villagers regarded it as a fabulous narration, inasmuch as the
reptile had never been seen but once, and then by only one person. The second time, however, had its effect, and the truth of
the matter the villagers resolved to settle. A detachment of young
men accordingly visited the locality, and by dividing their force
into squads they commenced the work of watching the hole out
of which the snake, if snake there was, had, in case he wished to
enjoy the beauties of the outer world, to emerge.

For three days and three nights—a Biblical coincidence in the
case of Jonah—did they keep their weary vigils, and disheartened,
they concluded that unless something should turn up or come
out by the evening of the fourth, to abandon the search and credit
themselves to being hoaxed individuals. But ere the allotted time
had waned, they were gratified by the appearance of the foul thing
which they had come in search of. Slowly it crept from its hiding
place, and as if aware that danger was lurking near, thrust out its
tongue in a defiant manner, and in a language peculiar to snakes,
hissed a warning that he was monarch of all he surveyed: but a
ball sped from the rifle of an unerring marksman, proved the
superiority of the human to the brute creation, and without a
moment's notice the monster was called upon to give up the ghost.
The excited detachment drew him forth, and judge of their surprise at finding him twenty-one feet in length. The old farmer's
story was verified, and the snake now dead and in captivity was
without doubt the same he had seen thirty years ago. It was found
to be a rattlesnake, and its extreme length from head to tail was
twenty-one feet six inches, and the greatest diameter of body
eighteen inches. It had one hundred and eleven rattles, which
would make its age one hundred and fourteen years, as they get a
rattle annually after attaining three years. The snake was con-

veyed to town, and has since been visited by tremendous crowds, anxious to satisfy themselves of what at first seemed to be an absurdity. Our informant further states that after being skinned and stuffed, it was eighteen feet three inches in length, and can be seen by and of the passengers on the above road, if they choose to step on the platform at the station.—Cincinnati *Commercial.*

Madison, Wisconsin, *Daily Wisconsin Patriot*
June 23, 1859

Indiana

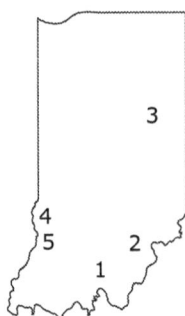

1. Crawford County
2. Scott County
3. Anderson, Madison County
4. Sullivan County
5. Knox County

A rattlesnake twelve feet long and eighteen inches in circumference has been seen in Crawford county.

Fort Wayne, Indiana, *Daily Gazette*
August 10, 1870

Latest Snake Story.
An Indiana Farmer Captures a Museum Attraction

Columbus, Ind., May 28.—James Graham, an old-time Virginian of fan-bark and hoop-pole notoriety, who resides in Scott county, was in this city, and tells a big snake story, about killing a monster rattlesnake unlike any ever before seen in Indiana. He says he was engaged in pulling bark, last week, and came in contact with a den of rattlers, and after killing ten large ones, as large as are generally seen in this locality, he started up the hillside, where a monster snake, which looked as large as all the balance, lay basking in the sun.

The rattler, on seeing him approach, set up his notes of warning. Graham retreated, and, procuring some hickory bark, made a lasso or harpoon by tying his bark spud to it and hurled it at his snakeship, striking the huge reptile back of the head. On measurement after death it was found to be nineteen feet in length, and had thirty-nine rattles and a button. The skin was stuffed and will be sent to a museum.

Marion, Ohio, *Daily Star*
May 28, 1891

Indiana's Big Snake
"Madison County Terror" Has Reappeared
Near the Dismal Swamp.

Near Anderson, Ind., the "Madison County Terror," probably the largest snake in Indiana, has made its appearance again. It was seen a few mornings ago just south of Anderson in the vicinity of the Dismal swamp, where it has made its home for the last three years. The snake is a monster and its origin has long been a mystery. Certain it is that it is not of Indiana origin.

Several years ago a circus train was wrecked in that vicinity of the swamp and several snakes got away. All were comparatively large. It is thought the reptile that has been making his annual appearance for three years is one of them. All of those snakes had their fangs taken out, and this probably accounts for the fact that the snake, which has lately been known as the

"Madison County Terror," does not often venture from his secluded and never-frequented retreat.

Farmer John Noland has often seen the reptile in the vicinity of his home. He describes it as being fully twenty feet in length and at the broadest point fully two feet in diameter. The track it leaves on a newly plowed corn field, says Noland, resembles the track that would be made by a person dragging a heavily filled wheat sack across the plowed ground.

Several times small animals have been missed in the vicinity of the swamp, and it has been thought that the snake, tiring of the food that it easily gets in the swamp, strolls out occasionally and gets some domestic meat. No one has ever been injured by the snake and it seems to avoid getting in close quarters.

A year ago it was in a wheat field near the Noland home when a binder was at work. It was badly scared and got out of the field before the hands could get at it. It was reported last fall that it had been killed. Those who saw it last week say it was just as big and cowardly as heretofore.

Syracuse, New York, *Herald*
July 18, 1897

Step High! "Big Jim" is in Indiana.
Monster Rattlesnake is Terrorizing the Agriculturists.

Indianapolis, Ind., August 15.— "Big Jim" a monster rattlesnake that has terrorized two generations of people in that part of Illinois just across the Wabash River from Knox and Sullivan counties, is said to have made his way into this state. He is now playing havoc on the farms of Knox and Sullivan county farmers.

The rattler is known as "Big Jim" because he scared a negro named Jim so badly 26 years ago that he fell into the Wabash river and was not seen again.

Wm. H. Thompson, a Sullivan county farmer, found "Big Jim" carrying away his chickens last night.

Thompson says the snake was ten feet long and fully five inches in diameter. The whir of his rattles "had a regular buzz-saw sound," and Thompson was afraid to shoot. The farmers of Knox and Sullivan counties are organized to do battle with the monster.

Elmer Holder, a farmer living near Dittney Hill, Warrick county, is said to have killed a chicken snake last night that was 16 feet long and eight inches in circumference. The same snake is said to have been a terror to the neighborhood for many years.

Lima, Ohio, *Daily News*
August 15, 1908

"Big Jim" of Indiana and Illinois is no more. He has gone to the "happy hunting ground" of snakedom. For twenty-six years he furnished material for all manner of snake" stories, and the veracious teller of the "dog day" yarns will now have to get new material.

"Big Jim" was a giant rattlesnake, four times the length of the ordinary rattlesnake, and twice as long as the diamond-back of the South. He is supposed to have come from Illinois, possibly driven out when towns went "dry" under local option. Several times during the summer he was seen in Sullivan and Knox counties, and his great size created terror wherever he roamed.

"Big Jim" became hungry, and, according to a Sullivan (Ind.) dispatch to the Indianapolis *News*, his hunger lead to his death. He wished to feast on pig, and a huge boar thwarted his desire. John Bascomb, an employe on the farm of W. H. Thompson, in the southwestern part of Sullivan county, heard a commotion near the pig pen, and when he reached the place he saw a boar fighting an enormous snake.

Exciting Fight With the Boar.

The boar had its mouth so fastened on the snake's head that its fangs could not be used. The giant reptile was thrashing its body backward and forward, attempting to get out of the clutches of its antagonist, but the boar held on, feet braced apart, while Bascomb ran to the farmhouse and got a 32-caliber revolver.

By the time Bascomb got back to the pen the other hogs there seemed to have discovered that something was wrong and were trying to help the big boar in his fight. They jumped on the body of the snake as it wriggled about and tried to stamp out its life. Bascomb entered the pen and got a good shot at the snake, breaking its neck. The shot caused the boar to drop the snake's head, but the snake died almost instantly. The boar had not been struck by the snake and is in good shape after its fight.

The snake was skinned and the skin measured. From tip to tip it is twelve feet five inches, and at its widest is fourteen and three-fourth inches across.

A Terror for Twenty-seven Years.

"Big Jim" that for the last twenty-seven years has been the talk and terror of that part of Illinois lying just across the Wabash river from Knox county, was thought to have crawled into Indiana and to be in the southwestern part of this county. He had not been seen in the haunts in the "snake rock" district in Illinois for several weeks, and meanwhile farmers in the northwest part of Knox county were reporting almost every day tales of a monster rattler that is having its own sweet way.

One of the last reports of the supposed "Big Jim" (so called because twenty-six years ago he scared a negro of a boat crew so badly that the negro, whose name was Jim, fell into the water and was never seen again) came when William H. Thompson, a farmer in the southwest part of the county, discovered a supposed "chicken thief" that had at this time more than 200 on his place. For a week he had missed young chickens every morning, sometimes as many as three being taken in the night.

His Snakeship a Chicken Thief.

Mr. Thompson decided to "lay" for the thief, and with his son, William H., jr., he waited near the coops, gun in hand and an electric pocket searchlight in his pocket, until 3 o'clock in the morning. He had heard no one approach but suddenly there was a great commotion in the coop yard, and in a minute all the chickens were squawking with fright. The Thompsons hurried into the yard and flashed the electric light. The father was prepared to shoot a man, but instead of a man he saw a huge snake swallowing a young pullet. At the same time the snake saw the men and glided away like a flash of light.

The snake was a rattler, fully ten feet long and five inches through, according to Thompson. When the reptile had crawled a little distance away, it stopped, coiled, and the whir of the rattles had a regular buzzsaw sound. The men did not try to pursue him, being frightened by the huge size of the reptile. They ran back into the house and listened at the windows. Two more pullets were taken from the chicken coops before the snake apparently satisfied its appetite and crawled away.

Syracuse, New York, *Herald*
August 30, 1908

Here Is Prize Snake Story.
Big Jim, the Monster Rattler and Terror of the Little Wabash, Paralyzes the Logging Business.

[Indianapolis *Star*]

O for another St. Patrick to rid the Little Wabash of its rattle-snakes! That is the prayer of river men, lumbermen and farmers, from Carmi to Shawneetown. It is in that section of Illinois and in the adjoining "pocket" district of Indiana that the prize snake story of the two states is told. It concerns "Big Jim," a monster rattler, and his reign of terror for more than a score of years. And the best narrator of this prize story is Captain Edward Ballard, a veteran river man of Carmi, across the river in Illinois.

Recently he told his story to the *Star*'s representative, and here it is:

In the early eighties, while saw-logging was good on the Skillet Fork, and when the Little Wabash was the mecca of river men in early spring, when high water facilitated their work of raft-ing the long line of logs down to the Ohio, an incident occurred that established the significance of snakes. This is no snake yarn, but is an actual occurrence. It can be vouched for by any of the old-time river men, and lots of the younger ones remember it.

"I was rafting a drift of logs down the stream early in May. We had just passed the mouth of Skillet Fork and had steered into the main current of the Little Wabash. The river was not so high as it had been earlier in the spring, and I had several men along to guard against jams. We had just got well into the stream when a blackened sky darkened our course. A heavy spring rain set in and we towed to shore. Tying our raft as securely as possible, we built a camp fire under a ledge of the rock on Rattlesnake Bluff. It had long been known to river men as an exceedingly danger-ous place to stop, as hundreds of thousands of rattlers were known to infest the locality. They wriggled there in force in the fall and found excellent quarters to hibernate in the dense clusters of rock that jab the shore of the stream.

"In my crew was a great hulking negro from Cincinnati. He was one of the largest [black men] I ever saw, and as strong as a mule. He was of a superstitious type, and we deemed it prudent not to tell him of the slimy inhabitants of the bluff. It was his first trip, and he knew but little of the country. The boys called him 'Big Jim,' and he was not incorrectly named.

"'Jim, go get some wood,' one of the boys ordered as the fire began to grow dim.

"'Mistah, it am mighty dark up dere. and it hain't bery co-o-o-ld nohow. I'se not wantin' to run up dem rocks fo' nuthin'. G'long yo'self.'

"He finally went. He was not more than half way up when one of my men yelled a word of warning to him about rattlesnakes. The night was inky dark save for an occasional flash of lightning, and was an ideal one for grewsome fears to seize the [man]. He had probably got half way up when we heard a startled cry. In the next instant there was a splash in the stream, as the frightened negro plunged in head first. He was praying for deliverance. We had but a few minutes to wait until we understood the cause of the commotion. Sounding his rattles to a martial time a big rattler was drumming for volunteers. In less than fifteen minutes the chorus was deafening, as the serpents rushed to the place of alarm.

"We jumped into a boat that we had tied to the raft of logs and rowed to the west side of the stream. We felt safe on the other side of the river, as there were no rocks there, and, as far as river men knew, no rattlers had ever been seen on that side. Rattlers do not like water, and are the poorest kind of swimmers. In fact they will never take to the water unless forced to. All night we could hear the din across the stream. There must nave been hundreds of snakes, judging from the noise they made. When day broke we went across to our logs, and several of the reptiles had crawled on them. They were quickly killed, and in a few hours we were ready to continue our journey. The bank was still lined with snakes, and among the number was one of the largest rattlers I ever saw. It must have been seven or eight feet long, and was proportionately large. We fired at it as we pulled out, but the shots went wild.

"We looked along the banks for 'Big Jim' as we went down, but found no trace of the missing man. We finally concluded that he had been so intensely frightened when he fell in the stream the previous night that he had made his way to the bank and left for Carmi, eight miles below. 'Big Jim' wasn't a lazy [man], and we finally settled on this conclusion as a reasonable solution of his disappearance.

"'What if the negro was drowned?' asked one of the crew as we pulled into Carmi.

"'There's no danger of that,' responded one of the men, 'because he's too good a swimmer for such a thing to happen.'

"We agreed that we would search the saloons for him, and search them we did. Our quest was in vain. No one had seen him. Had he indeed lost his life? Next day came and still nothing of 'Big Jim.' Farmers who lived near the bluff had not seen anything of him, and the story of his disappearance was causing a panic among the blacks all over the town. The news leaked out in a hurry, and finally the negroes got it that he had been devoured by the big rattler that was reputed to rule the bluff.

"Just to ease matters we sent some of the boys up to look for the negro's body. Nothing could be found of him and the men returned with the news. Consternation seized the colored 'river rats' and paralyzed the log business. Most of the laborers employed in skitting down the drifts had been of Ethiopian hue, but as the story of 'Big Jim's' fate spread the negroes grew restless when the name of Rattlesnake Bluff was mentioned.

"'No, sah, boss, I ain't gwine to go up to make bait fo' no snake and yo' all needn't think I'se gwine to change my min' about it. That big snake has done gone and gobbled Jim, and yo' all know it,' was the usual reply the negroes gave when asked to make trips up stream. Thousands of logs lay in the stream all year, and lots of the white people, too, became apprehensive of the bluff. A Cincinnati paper, I remember, got hold of the story and made a big write-up over the affair. The negro was pictured in the mouth of the big snake, and that was enough to send spasms of fear thrilling through the race.

"In the succeeding fall the big snake did do a stunt or two that forged his fame for all time to come. Scores of cattle and hogs died under circumstances indicating poisoning, and the blame was finally traced down to the rattler. Someone had called the snake in honor of the negro it drove away and it began to be known as 'Big Jim.' Of course it was always made out in the worst light possible. That winter Frank Sefried took a crew of men up to the bluff in an effort to dethrone the monster. It was argued that if the snakes were all driven away the negroes would go back to work in the spring. Dynamite, about enough of it to blast the Panama Canal, was used with telling effect, and rattlers of all sizes were blown from their rocky lair. But among them was not the form of 'Big Jim,' the ruler of the 'joint.' There were two approaches to the den under the main part of the bluff and both of these were destroyed with the explosive.

"Next spring a few of the more intrepid of the colored laborers were induced to go up stream. Things appeared to be getting along fine until about the first of June. One of the negroes saw the big snake perched on the bluff and swore that he had the missing negro's skull in his coils playing with it. The story had the effect we had worked so hard to dispel. Every negro as far down as Shawneetown heard of the affair and all of them steered clear of the place. It ruined the rafting business, which had just now got good again. That one fact saved the vast timber resources of the Skillet Fork from devastation, long ago, and to this day none of the old-time negroes will venture as far up stream as the bluff.

"William Ude, near whose farm the retreat of 'Big Jim' is located, has for over 20 years worn boots to guard against snake bites. Hundreds of rattlesnakes abound on the bluff, and last year Ude killed over 300 while breaking wheat ground.

"He is known to be over 25 years old and was fired upon by Lee L. Staley, the sheriff of White County in 1885. Up to date he has not been seen, but last fall James A. Welch, City Treasurer of Carmi, was almost scared out of his wits by the rattler.

"For years and years to come the people living in White County and vicinity will relate the stories of 'Big Jim,' the master of bluff and valley."

Washington, D. C., *Post*
April 21, 1921

Kentucky

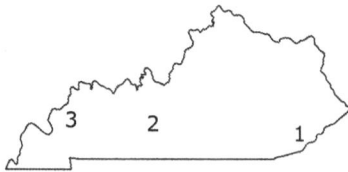

1. Harlem County, should be Harlan County
2. Rocky Hill, Edmonson County
3. Heath Mountain, Crittenden County

A Monster Reptile.

A correspondent of the Abington *Democrat,* writing from Walnut Hill, Lee county, Va. who is as the *Democrat* assures its readers, "a gentleman in whom implicit confidence may be placed," gives the following account of the killing of a monster reptile in Harlem county, Ky. He says:

About three weeks ago, five men went to gather whortleberries in the mountainous part of Harlem county, Ky.,—and in their travels came to a small branch at the foot of a steep ridge where they discovered a smooth beaten path or rather slide that led from the branch up to the ridge. Curiosity tempting them to know its meaning they followed the trail to the top of the ridge where, to their astonishment they found about an acre of ground perfectly smooth and destitute of vegetation near to the center of which they dis- covered a small sink or cave, large enough to admit a salt barrel.

They concluded to drop in a few stones, and presently their ears were saluted with a loud, rumbling sound, accompanied with a rattling noise; and an enormous serpent made his appearance, blow- ing and spreading his head, and his forked tongue protruding. The men were struck with affright, and suddenly the atmosphere was filled with a smell so nauseating that three out of the five were taken very sick; the other two discovering the position of their compan- ions, dragged them away from that abode of death: About ten- feet of the snake had, to their judgment made its appearance, when they hurried home and told to their neighbors what they had seen.

The next day were mounted some ten of the hardy mountain- eers armed with rifles, determined to destroy the monster. On approaching within 100 yards of the dwelling of his snakeship their horses suddenly became restive and neither force nor kindness could make them go any nearer.—The men dismounted and hitching their horses proceeded on foot with rifles cocked to the mouth of the cave. They hurled in three or four large stones and fell back some fifteen steps when the same noise was heard as before and out came the dreaded reptile, ready, as his looks indicated, to crush the intruders.

About the same length of the snake had appeared from the hole when eight or ten bullets went through his head and, as the monster died, he kept crawling out until 20 feet of that huge boa lay motionless on the ground. It was a rattlesnake with 28 rattles— the first was four inches in diameter, the rest decreasing in size to the last. With difficulty the men dragged him home and his skin can now be seen by the curious, in Harlem county.

Weyauwega, Wisconsin, Weyauwegian
September 18, 1857

A Kentucky Rattler.

[*Courier-Journal* Rocky Hill Special]
Great excitement prevailed in town the other evening over the capture of a huge rattlesnake While Dr. J B. Thomas was cutting wheat on his farm, adjoining the suburbs of town, himself and

hands heard a shrill rattling noise near by. Going in the direction of the sound the form of a huge snake could be seen. They at once prepared to capture the monster, which they finally succeeded in doing. They at once carried it to Rocky Hill and placed it in a secure place. The huge monster measures nine feet seven inches in length and fourteen inches in circumference.

Drs. Thomas and Rose extracted its teeth, and placed them in alcohol for exhibition. They measured one and a half inches in length. The snake is now on exhibition at L. M. Ewing's drug store. Mr. Ewing has had its picture taken and sent to the druggist at Smith's Grove to be seen by the excited people.

Defiance, Ohio, *Democrat*
June 28, 1883

Giant Rattler

Nobe Bean of Crittenden reports a giant rattlesnake in the Heath Mountain section which is about the biggest thing of the season. Bean has been busy inventing plans to capture alive the monstrous snake, but has not yet succeeded, although he has got close enough to the ugly reptile to estimate its dimensions, which are about the circumference of the average telephone post and length of some twelve feet, with rattlers about a foot long. Bean says he can sell the big snake to a circus for $1,000 if he can deliver it alive.—Sturgis (Ky.) *News-Democrat*.

New York, New York, *Times*
February 3, 1917

Louisiana

1. Junction City, Claiborne/Union Parishes

Bones of 100 Year Old Snake Found in Cave

Monroe, La.—The skeleton of a huge snake, measuring ten feet long and well preserved, and believed to have been more than 100 years old when it died, was found recently by a party of engineers and surveyors in a cave near Junction City, on the Louisiana-Arkansas state line. When the skeleton was found an aged negro living near the cave told members of the party a monster rattlesnake had terrorized the Junction City section sixty years ago.

La Crosse, Wisconsin, *Tribune*
August 14, 1923

Missouri

1. Independence, Jackson County

Huge Snake Story
A Rattlesnake Weighing One Hundred and Twenty-Five Pounds

From the Kansas City *Times*.

The following may, perhaps, be considered by many readers of the *Times* as a huge snake story. A responsible citizen of this city, however, gave it to a *Times* reporter as an actual fact, and although the reporter was at first loth to believe it, he soon became convinced that such was the fact. Mr. Snodgrass, who for many years was a grocery merchant of this city, but who has for a number of years past resided about ten miles east of

Independence, was out on his farm a few days ago, when he saw a huge serpent coiled up in a fence corner. Not daring to attack it with an ordinary weapon, he went to his house and, loading his shot-gun with a heavy charge of buck-shot, he returned to the place where the enemy of mankind still held the fort, and opening fire, succeeded in killing it. The dead serpent was weighed and pulled the scales at one hundred and twenty-five pounds, and had forty-two rattles attached to it. This is undoubtedly the largest serpent ever killed in this section of the country, and may be classed among the boa constrictor tribe. The skin of the reptile has been stuffed, and will be on exhibition at our forthcoming Exposition.

Ottawa, Kansas, *Journal Triumph*
August 29, 1878

Oklahoma

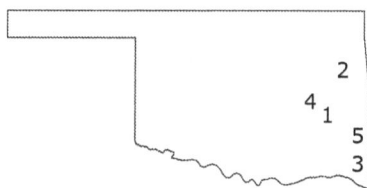

1. Eufaula, McIntosh County
2. Tahlequah, Cherokee County
3. Broken Bow, McCurtain County
4. Henryetta, Okmulgee County
5. Poteau, Le Flore County

Dobie (1982) noted an old story: "The Dallas *Weekly Herald* of July 28, 1877, reported a rattler killed in the Cherokee Nation, near Eufaula, eighteen feet long, with 'thirty-seven distinct rattles on its tail.'"

A Monster Rattlesnake.

The oldest inhabitant cannot remember a rattlesnake to compare with one just unearthed in the southern part of the Cherokee nation, says a letter from Tahlequah. The discovery came about through the construction work on the Arkansas

Valley railroad. A gang had been engaged for several weeks in excavating a deep cut in the mountains about fifteen miles southwest of Tahlequah, and has been from time to time disturbed and killed rattlers of quite respectable dimensions. The other day, however, there occurred an event in snake-killing which eclipsed all former exploits, and which excited even the stoical Cherokees. A big charge of giant powder was fired, and as the masses of rock came tumbling down they revealed the writhing folds of a monster rattlesnake. Two other reptiles were dislodged from their rocky retreat, but were little fellows compared with the other. They were about six feet long and rattled away noisily with their eight and ten joints. But above the din they made arose the deep, sonorous music of the monster. The sounds, in comparison, were like the snare and the bass drum. The two smaller snakes were lively and gave the railroaders a good fight, but the big one was sluggish and seemed dazed. He was easily approached, and was killed without much mutilation. After the dimensions had been taken and the rattles carefully counted the skin was stripped off with a view to stuffing. The big snake measured 12 feet 9 3/4 inches in length. The body was as large as that of a man of average size. The unusual number of twenty-four rattles bore testimony to the age of the former as well as to the quietness of the life he had led. Several of these rattles measured two inches across.

Marshfield, Wisconsin, *Times*
December 30, 1887

Champion Snake Story

A big rattlesnake is reported to have been killed by the graders on the Midland Valley road, south of Broken Bow. The snake is said to have been nineteen feet in length, weighed 127 pounds and, its rattlebox contained fifty-four rattles. It is said this specimen was killed by an explosion of powder or dynamite which was put in a stone bluff to tear it down so the roadbed might be graded to the proper level, and the snake was in a crevice among the rocks. It was skinned and cut open. One shoat and four jackrabbits were found in its interior.—Topeka (Kas.) *Journal.*

Altoona, Pennsylvania, *Mirror*
September 29, 1904.

Giant Snake Seen

Henryetta, Okla.(AP)—A rattlesnake, described as big around as a log, has been reported seen several times southeast of here. Lloyd Ingram said his son and some companions almost ran their car over the reptile on a country road.

Amarillo, Texas, *Globe-Times*
August 6, 1957

Klauber (1984) noted an early report of a 21-foot long rattlesnake that was killed near Poteau, Oklahoma.

Texas

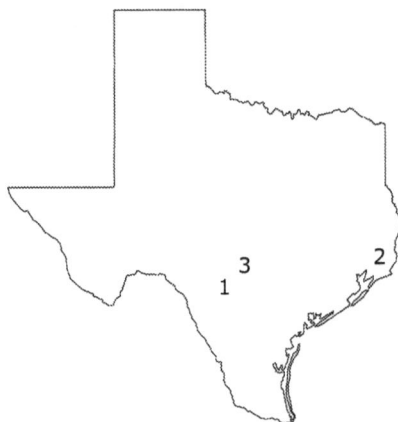

1. Quihi, Medina County
2. Hardin County
3. Rudolph, Kendall County

Dobie (1982) noted an early report, published in 1858: "Near Quihi, now in Medina County, the Abbé relates, 'a tiger hunter killed a rattlesnake he had mistaken for a dead tree' fallen to the ground. ... 'The reptile measured seventeen feet in length, eighteen inches in circumference and had twenty-five rings, or rattles.'"

Monster Rattlesnake.
It Was as Big as a Pine Log and Carried Eighty-One Rattles.
Special to the *News*.

Olive, Hardin Co., Tex., June 10.—While out in the woods inspecting timber lands, Mr. V. A. Fetty and G. A. Sternenberg, together with their woods foreman, Mr. E. P. McDonald, rode upon a rattlesnake, which they believe to be the oldest and largest in East Texas. They discovered it at some distance and being very anxious to become in possession of its large string of rattle soon began battle with the reptile. They soon won the fight and are now wearing the belt of 81 rattles to prove their championship. The snake after being killed had the appearance of a small pine log as it lay its full length—12 feet 6 inches.

Galveston, Texas, *Daily News*
June 12, 1902

Section Workers Kill Big Rattler
Giant Texas Reptile Weighs 113 Pounds.

Raymondville, Texas.—Section hands at Rudolph, a station on the Missouri Pacific lines to Kendall county north of here, killed a rattlesnake which weighted 113 pounds, according to the section foreman, Will Reeves, who was in Raymondville.

Reeves stated that the rattlesnake had 24 rattles and measured four inches across the head. He said the monster snake crawled right into the midst of the section crew, before they were aware of its presence, and began striking at the men, all the time making a hideous whirring noise with his rattles. He was 11 feet long, Reeves stated.

The men killed the reptile by plunging crowbars into his body, after the manner of throwing spears.

Reeves states that he has lived and worked on the Missouri Pacific line as section foreman in Kendall county ever since its construction through those wilds, and that he has killed rattlers with as many as 40 rattles, but that it was the largest one that he had ever seen.

Wellsboro, Pennsylvania, *Gazette*
August 4, 1927

Utah

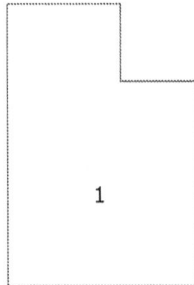

1. Salina, Sevier County

The Prize Snake Story
A Salina man Claim to Have Seen a Mountain Monster

The father of all rattlesnakes has at last been discovered. The reptile is making its home on the big hill near Rex's ranch some eight or ten miles southeast of Salina. His snakeship was seen about ten days ago and from the description given he must certainly be a whooper. A Salina man and his wife were driving over the hill at the time mentioned and when near the summit, their horse abruptly stopped and no amount of coaxing or abuse served to urge him forward. There was a strange foetid smell in the air and the gentleman remarked to his wife that some kind of a wild animal was concealed in the heavy undergrowth. He was about to alight from the buggy and make an investigation when a peculiar rustling sound in the road ahead caused him to pause. A moment later a gigantic head parted the brush and an enormous serpent glided out into the road. When the worthy couple surveyed the monstrous outlines of the snake they were stricken dumb with terror. The horse fairly squealed with fright and this seemed to arouse the ire of the snake. The big ophidian shook itself together in a multiplicity of sinuous folds and the bright markings on its lithe body glistened like bars of silver in the sunlight. Then the reptile touched the button and its tail did the rest. The sides of the mountains echoed the awful music and big stones and boulders became detached from their fastenings and crashed into the road. The old horse simply went wild and his owner gave him the lines. How they reached the foot of the hill without an awful accident neither of the occupants of the vehicle seem to know. Sufficient time has elapsed to secure actual facts concerning this giant serpent. The gentleman who saw it declares that the snake was fully seventeen feet in length, as big around the body as a nail keg and that it had a head like a crocodile. He counted its rattles. There were one hundred and sixty three besides the regulation button.—Salina *Press*.

Utah Journal
October 4, 1893

West Virginia

1. Logan County

Roving the Valley
With Friends and Neighbors
by Adrian Gwin

Absolutely nobody will believe this one.

It's another story like the old man from the hills who went to the Cincinnati zoo and saw his first giraffe.

He gazed at the tall monstrosity for some minutes, open-mouthed. Then he stamped his blackthorn walking stick and shuffled off mumbling.

"It's a dern lie! There ain't no such animal!"

We did almost the same thing when we went into the little tavern in Hamlin. "Go over to The Smokehouse and see the big rattlesnake hide they got tacked to the wall," Charlie Plumley told us.

"Biggest rattlesnake ever killed in the world!"

We went to see. (The Encyclopedia Britannica says "the largest of the rattlesnakes is the diamondback, which may reach a length of eight feet".)

There was the hide, stretched and crackly-dry and old, nailed to the wall of the Smokehouse tavern.

Bernie Skeens Killed It

The head is of course gone. The rattles (there were 28) are gone. But from one end to the other, the shrunken dry hide of the black diamondback rattler measures 12 feet, seven inches!

"It was a mite over 14 feet long when Bernie Skeens killed it," said Crawford Plumley, the tavern keeper.

"Long time ago, about 18 years ago, down in Logan county, I knew Bernie well and was an old buddy of his. He and some other fellow, I forget who, were squirrel hunting up Davis Run of Bandmill hollow above Logan. Bernie died about four years ago. Wish he was still alive to tell it himself.

"He came on this thing all coiled up, in a little cove underneath a rock ledge, right beside the path. He shot it three times. See the holes and ragged places there in the hide?

Won His Bet?

"Then the fool cut off the rattles and left the snake there. He went down to a beer joint near the mouth of Rum creek and started blowing about the 'eight or ten foot' rattlesnake he'd killed.

"Somebody bet him $10 it wouldn't come to 10 feet, so they went back and drug that thing down the mountain to the beer garden and there's where it was skinned out."

For identification purposes Crawford has tacked the hide of a small black diamond rattler (only 42 inches long) beneath the hide of the giant snake. The markings on both are the same, though of course the old hide is darkened with age and dust.

"Fellow from Huntington recently said he'd have one of the Marshall college professors come here and take a little sample of the hide, to run tests and prove if it is a real rattlesnake skin or not. He hasn't showed up yet, but I wish he would!"

John Shanholtz, a mechanic eating lunch at the Smokehouse, spoke up from another booth, "I've been called a liar many a time over that snake skin! Send me one a them pictures, will ya?"

Mate Still Living?

Crawford resumed his beer and sandwich, and continued, "They claim around down in Logan that there's the mate to this one still there in the big rock cliff.

"Some claim they've seen it. Say it's too rocky and steep to try to capture the snake, too dangerous, Why, a fellow that size would knock you down and kill you when it struck!

"People come here from all over just to see that snake skin. None of 'em believes it after they see it, though," he said sadly, munching his sandwich.

"Old man come here a while back, stood there a little bit, and then turned and walked out. 'That's an alligator skin!' he says, and he left.

"Now there's a man over West Hamlin who either saw that snake before it was skinned out, or saw the hide right afterwards, I can't recall which. He's a fellow named Tompkins, runs a grocery store over there. In fact, he might a-been workin' at that beer garden when Bernie drug the snake down off the mountain."

You Can See It, Too

We didn't have time to look up Mr. Tompkins, though it would have been interesting to know more about the biggest rattlesnake skin we've ever seen, anywhere.

We stepped out on the sidewalk and it came to the mind that there just ain't no such rattlesnake in the world. But there is the hide, 12 feet, seven inches long, 15 inches across at the middle.

You go by the Smokehouse there on Lynn street, right on Rt. 3 as you go through Hamlin, and you can go in and see it yourself.

But you won't believe it either. Nobody ever does.

Charleston, West Virginia, *Daily Mail*
October 1, 1954

My attempts to locate this specimen have not panned out. The Smokehouse is long gone. Might it have been a boa skin (given the lack of head and rattles)? Maybe, but the mention of a similarity in pattern to a "normal" timber rattlesnake suggests otherwise. This case is worth some effort, if anyone in West Virginia has the time.

Conclusion

We do not have enough data to show that any undescribed species or variety of giant snake exists in North America. What we do have, though, is enough data to point to certain geographic areas that could be searched for undescribed species or varieties of giant snakes.

This could be winnowed down further (and more accurately) by someone with a good knowledge of statistics and ecological modeling. Ecological modeling has been the subject of many recent studies, as it offers a way to predict geographic locations for rare or even unknown species (Pearson, et al, 2007; Raxworthy, et al, 2003).

As far as fieldwork is concerned, the one good thing about hunting mystery snakes is that they can be sought and captured just like any other snake. The equipment is pretty much the same, even with modifications. Any good herpetological field guide or textbook should cover tools and traps used to catch snakes. Now, there will be trial and error, as different species of snakes require different tactics in order to maximize results. You have to hunt them at the right time, the right season, and under the right conditions (Hoyer 2007). It will take effort, and time; not just a two-week vacation to a suspected locale. Until a species is confirmed, and its habits thoroughly studied, it is highly unlikely that a hit-or-miss approach will be immediately fruitful.

In order to confirm a new species or variety of snake, a specimen is required. This is especially important for a new species. The importance of a voucher is not just to confirm identity, or even to support legislative protection. Future conservation of a species requires knowing for certain that an animal exists, so that decisions regarding development and environmental impact are fully informed as to possible consequences. Today, it isn't enough just to pass a feel-good law allegedly protecting a species. We can't remain that naive.

385

Resources

Arment, Chad. 1995. Giant snake stories in Maryland. *INFO Journal* (73): 15-16.

Arment, Chad. 2004. *Cryptozoology: Science & Speculation*. Landisville, PA: Coachwhip Publications.

Arment, Chad. 2006. *The Historical Bigfoot*. Landisville, PA: Coachwhip Publications.

Ashton, Kyle G. 2001. Body size variation among mainland populations of the western rattlesnake (*Crotalus viridis*). *Evolution* 55(12): 2523-2533.

Ashton, Kyle G., and Chris R. Feldman. 2003. Bergmann's Rule in nonavian reptiles: turtles follow it, lizards and snakes reverse it. *Evolution* 57(5): 1151-1163.

Aubret, Fabien, and Richard Shine. 2007. Rapid prey-induced shift in body size in an isolated snake population (*Notechis scutatus*, Elapidae). *Austral Ecology* 32(8): 889-899.

Ayers, D. Y., and R. Shine. 1997. Thermal influences on foraging ability: body size, posture and cooling rate of an ambush predator, the python *Morelia spilota*. *Functional Ecology* 11: 342-347.

Baughman, Jon. 1987. *Strange and Amazing Stories of Raystown Country*. Saxton, PA: Broad Top Bulletin.

Baughman, Jon, and Ron Morgan. 1977. *Tales of the Broad Top*. Saxton, PA: Broad Top Bulletin.

Boback, Scott M. 2003. Body size evolution in snakes: evidence from island populations. *Copeia* 2003(1): 81-94.

Boundy, Jeff. 1995. Maximum lengths of North American snakes. *Bulletin of the Chicago Herpetological Society* 30(6): 109-122.

CNAH (Center for North American Herpetology). 1994-2007. http://www.cnah.org/nameslist.asp?id=6

Christman, Steven P. 1975. The status of the extinct rattlesnake, *Crotalus giganteus*. *Copeia* 1975(1): 43-47.
Coleman, Loren. 2001. *Mysterious America: The Revised Edition*. New York: Paraview Press.
Conant, Roger, and Joseph T. Collins. 1998. *A Field Guide to Reptiles and Amphibians: Eastern and Central North America*. 3rd edition. Boston: Houghton Mifflin.
Devitt, Tom, et al. 2007. *Pituophis catenifer* (Bullsnake). Maximum length. *Herpetological Review* 38(2): 209-210.
Dobie, J. Frank. 1982. *Rattlesnakes*. Austin, TX: Univ. of Texas Press.
Ferriter, Amy, et al. 2006. Chapter 9: The status of nonindegenous species in the South Florida environment. 2006 South Florida Environmental Report. Florida Department of Environmental Protection / South Florida Water Management District.
Fitch, Henry S. 2006. Gopher snakes, bullsnakes and pine snakes. *Journal of Kansas Herpetology* 17: 16-17.
Gray, P. L. 1905. *Gray's Doniphan County History: A Record of the Happenings of Half a Hundred Years*. Bendena, Kansas: Roycroft Press.
Holman, J. A. 2000. *Fossil Snakes of North America*. Bloomington, IN: Indiana University Press.
Hoyer, Richard F. 2007. The fallacy of perceptions. *Journal of Kansas Herpetology* 21(March): 5-10.
Jessop, Tim S., et al. 2006. Maximum body size among insular Komodo dragon populations covaries with large prey density. *Oikos* 112(2): 422-429.
Keel, John A. 1994. *The Complete Guide to Mysterious Beings*. New York: Doubleday.
Keogh, J. Scott, Ian A. W. Scott, and Christine Hayes. 2005. Rapid and repeated origin of insular gigantism and dwarfism in Australian tiger snakes. *Evolution* 59(1): 226-233.
Klauber, Laurence M. 1984. *Rattlesnakes: Their Habits, Life Histories, and Influence on Mankind*. Abridged edition. Berkeley, CA: University of California Press.
Madsen, Thomas, and Richard Shine. 2000. Silver spoons and snake body sizes: prey availability early in life influences long-term growth rates of free-ranging pythons. *Journal of Animal Ecology* 69: 952-958.
Madsen, Thomas, and Richard Shine. 2002. Short and chubby or long and slim? Food intake, growth and body condition in free-ranging pythons. *Austral Ecology* 27: 672-680.

Makarieva, Anastassia M., Victor G. Gorshkov, and Bai-Lian Li. 2005. Gigantism, temperature and metabolic rate in terrestrial poikilotherms. *Proceedings of the Royal Society* B 272: 2325-2328.

Manthey, U., and W. Grossmann. 1997. *Amphibien & Reptilien Südostasiens*. Münster: Natur und Tier-Verlag.

Maretiæ, Z., and F. E. Russell. 1979. An unusual nonvenomous snake bite. *Toxicon* 17: 425-427.

Meshaka, Jr., Walter E. 2006. An update on the list of Florida's exotic amphibian and reptile species. *Journal of Kansas Herpetology* (19): 16-17.

Mori, Akira, and Masami Hasegawa. 2002. Early growth of *Elaphe quadrivirgata* from an insular gigantic population. *Current Herpetology* 21(1): 43-50.

Moseley, E. L. 1901. The python in Pennsylvania. *Science* 14(361): 852-853.

Naish, Darren. 2007. Stupidly large snakes, the story so far. *Tetrapod Zoology* (blog). May 31, 2007. http://scienceblogs.com/tetrapodzoology/2007/05/stupidly_large_snakes_the_stor.php

Olalla-Tárraga, Miguel Á., Miguel Á. Rodríguez, and Bradford A. Hawkins. 2006. Broad-scale patterns of body size in squamate reptiles of Europe and North America. *Journal of Biogeography* 33: 781-793.

Pearson, Richard G., et al. 2007. Predicting species distributions from small numbers of occurrence records: a test case using cryptic geckos in Madagascar. *Journal of Biogeography* 34(1): 102-117.

Raxworthy, Christopher J., et al. 2003. Predicting distributions of known and unknown reptile species in Madagascar. *Nature* 426(Dec. 18-25): 837-841.

Reed, Robert N. 2005. An ecological risk assessment of nonnative boas and pythons as potentially invasive species in the United States. *Risk Analysis* 25(3): 753-766.

Reed, Robert N., and Scott M. Boback. 2002. Does body size predict dates of species description among North American and Australian reptiles and amphibians? *Global Ecology & Biogeography* 11: 41-47.

Rumer, Tom. 1999. *Unearthing the Land: The Story of Ohio's Scioto Marsh*. Akron, OH: University of Akron Press.

Skitt. 1859. *Fisher's River (North Carolina) Scenes and Characters*. New York: Harper & Bros.

Smith, James M. 2004. *Bigfoot and Other Mystery Animals of Alabama*. Wadley, AL: James M. Smith.

Stebbins, Robert C. 1985. *A Field Guide to Western Reptiles and Amphibians*. 2nd edition. Boston: Houghton Mifflin.

Stephens, Lee. 1988. Slithering snakes and snallygasters! Strange creatures may be wandering through southern Maryland. *Maryland Independent.* (October 28, 1988)

Weatherhead, Patrick J., and Gabriel Blouin-Demars. 2003. Seasonal and Prey-Size Dietary Patterns of Black Ratsnakes (*Elaphe obsoleta obsoleta*). *American Midland Naturalist* 150: 275-281.

Webb, Nancy McDaniel. 2002. The Giant Snake. http://archiver.rootsweb.com/th/read/ALMARION/2002-06/1023932047

Willoughby, Hugh L. 1910. *Across the Everglades.* Philadelphia: J. B. Lippincott Co.

Coachwhip Publications

CoachwhipBooks.com

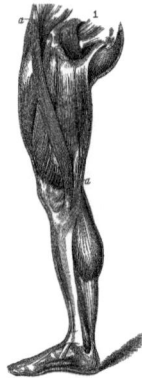

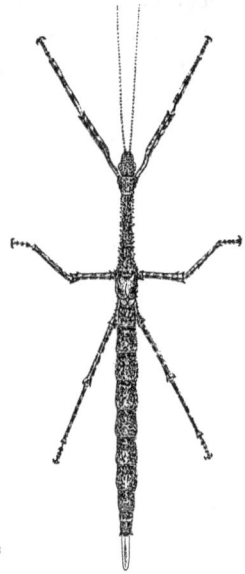

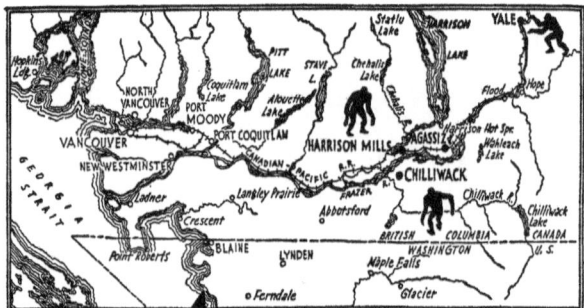

Coachwhip Publications

CoachwhipBooks.com

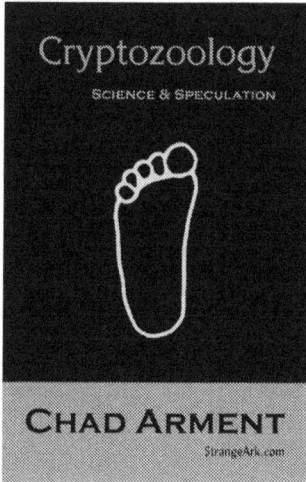

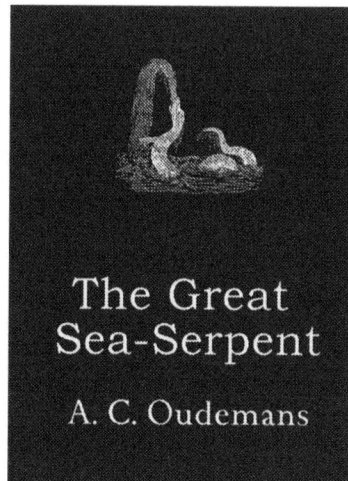

Cryptozoology
SCIENCE & SPECULATION

CHAD ARMENT
StrangeArk.com

The
Historical
Bigfoot

RANCHERS BEGIN
WILD MAN HUNT
ALONG ROCKIES

Prospectors Say
Wild Man Rules
Arctic Kingdom

TERRIBLE SASQUATCH
ABROAD IN THE LAND

Chad Arment

Cavemen Roam the Rockies?

Sea Monsters
Unmasked

Henry Lee

The Great
Sea-Serpent

A. C. Oudemans

www.ingramcontent.com/pod-product-compliance
Lightning Source LLC
Chambersburg PA
CBHW020602270326
41927CB00005B/134